D1647859

# Communication and Computer Networks

# COMMUNICATION AND COMPUTER NETWORKS

## Modelling with discrete-time queues

**Michael E. Woodward**
*Loughborough University of Technology*

IEEE Computer Society Press
Los Alamitos, California

Washington • Brussels • Tokyo

Published by the
IEEE Computer Society Press
10662 Los Vaqueros Circle
P.O. Box 3014
Los Alamitos, CA 90720-1264



IEEE Computer Society Press Order Number 5170-04
ISBN 0-8186-5171-7 (microfiche)
ISBN 0-8186-5172-5 (case)

Library of Congress Cataloging in Publication Data available



Additional copies can be ordered from

IEEE Computer Society Press
Customer Service Center
10662 Los Vaqueros Circle
P.O. Box 3014
Los Alamitos, CA 90720-1264
Tel: (714) 821-8380
Fax: (714) 821-4641
Email: cs.books@computer.org

IEEE Service Center
445 Hoes Lane
P.O. Box 1331
Piscataway, NJ 08855-1331
Tel: (908) 981-1393
Fax: (908) 981-9667

IEEE Computer Society
13, avenue de l'Aquilon
B-1200 Brussels
BELGIUM
Tel: +32-2-770-2198
Fax: +32-2-770-8505

IEEE Computer Society
Ooshima Building
2-19-1 Minami-Aoyama
Minato-ku, Tokyo 107
JAPAN
Tel: +81-3-3408-3118
Fax: +81-3-3408-3553

The Institute of Electrical and Electronics Engineers, Inc.

Printed in Great Britain by Bookcraft (Bath) Ltd.

# Preface

In the early 1970s something of a revolution was taking place in the world of signal processing. This was mainly due to the introduction of digital signal processing devices which were rapidly replacing the old analogue ones. In parallel with this came the introduction of new, discrete-time techniques which could be used for the analysis and design of the new digital systems. The use of the $z$-transform and the fast Fourier transform (FFT) algorithms and techniques based on these became widespread.

A similar revolution was also taking place in the early 1970s in the computer and telecommunications industries, with the introduction of microcomputers, and computer networks such as the Aloha system and the ARPA network. However, this is where the parallel with signal processing ends. Despite the digital nature of these computer systems and networks, the techniques used for their analysis and design were firmly embedded in queuing theory, a topic that was mainly exclusive to the continuous-time domain. This is a situation that has persisted almost to the present day, and it is only now, in the 1990s, that the concept of modelling computer or telecommunication systems using discrete-time queuing theory is beginning to generate considerable interest. It is in anticipation of this interest escalating that this book has been written.

The reluctance to use discrete-time concepts is mainly due to the generally accepted view that discrete-time queueing systems can be more complex to analyse than equivalent continuous-time ones. In a continuous-time model, only a single state change can occur at any given time instant. In a discrete-time model, because of the finite size of a time-unit, multiple state changes can occur from one time-unit to the next. This complicates the resulting analysis of the model.

However, many digital systems such as communication or computer networks operate on the basis of time-slotting. Such systems inherently lead themselves to representation with discrete-time models by specifying a one-to-one correspondence between a time-unit in the model and a time-slot (or some convenient multiple thereof) in the physical system. In this way very accurate models, that faithfully reproduce the stochastic behaviour of a communication or computer network, can be constructed. One of the objectives of the present book is to present a unified approach to developing accurate models of communication or computer networks using discrete-time queueing theory.

Although the models so constructed cannot always be solved exactly, another objective of the book is to develop good approximation techniques to obtain solutions.

To motivate the above two objectives, I quote the words of Paul Schweitzer when taking part in a panel discussion on 'The State of the Art and Future Directions in Computer Modelling,' the text of which can be found in [IAZE 84]:

> 'In my view we have reached the end of the road for exact models, and future efforts should be devoted to developing better classes of approximation models ... it is better to have an approximate treatment of an accurate model than an accurate treatment of an inaccurate model'.

This book attempts to marry these two suggestions, and provide a unified method for developing accurate, discrete-time models of communication and computer networks, along with good approximation techniques for solving them.

The book has seven chapters and, following the Introduction (Chapter 1), can be divided into three parts which might be classed as: Basics (Chapters 2 and 3), Discrete-time queueing systems (Chapters 4 and 5), and Applications (Chapters 6 and 7).

Chapter 1 outlines the basic aims and philosophy of the book, and introduces discrete-time queues at the simplest possible level.

Chapters 2 and 3 cover probability theory and discrete-time Markov chains. It is expected that most readers already have a knowledge of these topics. The topics are, however, included for completeness, to establish a notation, and provide a reference source for subsequent chapters. The chapter on probability theory (Chapter 2) is fairly standard, with the possible exception of the material on the conditional binomial distribution, which later turns up in one of the models. Surprisingly, this material does not appear in the popular texts, such as those by Feller [FELL 71] or Papoulis [PAPO 84]. Chapter 3 on Markov chains is another standard treatment. There is however a strong emphasis here on how Markov chains relate to performance models for communication and computer networks. There is also an emphasis on the concept of time reversal.

Chapters 4 and 5 deal with discrete-time queues and discrete-time queueing networks, respectively. Although many of the results presented in Chapter 4 can be found elsewhere, these are spread over a number of diverse research papers that use widely different notations. I have derived the results from basics, mainly using nothing more complicated than discrete-time Markov chains, in order to give a consistent presentation throughout. Of particular importance here are the results on *S*-queues, which are based on a paper by Jean Walrand [WALR 83A].

Chapter 5 extends the results on discrete-time queues to discrete-time queueing networks. Again there is a strong influence from Walrand here. The classic product form distributions are derived for both open and closed networks of *S*-queues. A new development here is the discrete-time queueing network model for multiple-access protocols. In general, these models have state dependent routing probabilities and non-linear traffic equations, and approximation techniques must be used to solve them. An approximation technique known as equilibrium point analysis (EPA) is introduced. This method was motivated by the work of Shuji Tasaka [TASA 86]. The emphasis in this text however is on the interpretation of EPA in the context of discrete-time queueing networks. Important extensions of the method are derived that can be applied to networks with different customer classes.

Chapters 6 and 7 are concerned with applications to satellite networks and local networks, respectively. Most of the models here have been carefully chosen to illustrate a specific modelling concept or technique; for example, how to handle such things as buffered users, statistically different users, finite channel delays, timing deadlines (as in real-time), and unsolvable traffic equations are all covered. A number of the models presented here have not previously been published in the open literature.

The precedence relations among the chapters are strictly sequential, although, as previously noted, Chapters 2 and 3 can mostly be skipped by those familiar with the topics. Some exercises are included at the end of Chapters 3 through 7. These provide plausible extensions of the results contained in the text, although those exercises marked with an asterisk are considered to be fairly major undertakings.

The text should prove useful to both practitioners and researchers concerned with communication and computer network performance evaluation. It is hoped that it will also prove invaluable to 'get off the ground' prospective researchers who wish to investigate the relatively untapped area of discrete-time queueing systems and their application to the performance modelling of communication and computer networks.

**Acknowledgement**

I would like to thank colleagues and students in the Department of Electronic and Electrical Engineering at Loughborough University of Technology for many helpful suggestions, and also acknowledge the use of the facilities at the University. Special thanks are due to my friend and student Kamajith (Rohan) Rodrigo who expertly typed the manuscript, produced most of the diagrams, and corrected many of my errors along the way. Thanks are also due to Professors Jean Walrand and Isi Mitrani, both of whom helped to clarify points on which I was unsure. Finally, I thank my wife Christine, without whose support and understanding this book could not have been written.

Michael E. Woodward
Loughborough

# Contents


| | | |
|---|---|---|
| **1** | **NETWORKS, QUEUES AND PERFORMANCE MODELLING** | 1 |
| | 1.1 Introduction | 1 |
| | 1.2 Network types | 2 |
| | 1.3 Multiple-access protocols | 3 |
| | 1.4 Discrete-time queues | 6 |
| | 1.5 Performance measures | 10 |
| **2** | **PROBABILITY, RANDOM VARIABLES, AND DISTRIBUTIONS** | 12 |
| | 2.1 Probability | 12 |
| | 2.2 Random variables | 19 |
| | 2.2.1 Moments of a random variable | 24 |
| | 2.2.2 Generating functions | 27 |
| | 2.3 Distributions | 33 |
| | 2.3.1 The uniform random variable | 33 |
| | 2.3.2 The exponential random variable | 37 |
| | 2.3.3 The geometric distribution | 39 |
| | 2.3.4 The binomial distribution | 42 |
| | 2.3.5 The Poisson distribution | 42 |
| | 2.3.6 The Normal distribution | 44 |
| | 2.4 Conditional distributions | 45 |
| **3** | **STOCHASTIC PROCESSES AND MARKOV CHAINS** | 49 |
| | 3.1 Poisson process | 49 |
| | 3.2 Properties of the Poisson process | 51 |
| | 3.2.1 The union property | 51 |
| | 3.2.2 Decomposition property | 52 |
| | 3.2.3 Interarrival times | 53 |
| | 3.3 Markov chains | 53 |
| | 3.3.1 Definitions and terminology | 54 |
| | 3.3.2 Equilibrium distribution | 57 |
| | 3.3.3 Reversible chains | 63 |
| | 3.4 Markov chain models | 69 |
| | 3.5 Exercises | 74 |
| **4** | **DISCRETE-TIME QUEUES** | 76 |
| | 4.1 Performance measures and Little's result | 77 |
| | 4.1.1 Performance measures | 77 |
| | 4.1.2 Little's result | 77 |
| | 4.2 Discrete-time queueing conventions | 78 |

4.3 Discrete-time $M/M/1$ queue 79
4.4 Discrete-time $M/M/1/J$ queue 83
4.5 Discrete-time $M^{a_n}/M/1$ queue 85
4.6 Discrete-time $M^{a_n}/M^{d_n}/\infty$ queue 90
4.7 $S$-queues 93
4.8 Exercises 98

**5 DISCRETE-TIME QUEUEING NETWORKS** 99
5.1 Tandem $S$-queues 100
5.2 Networks of $S$-queues 101
5.3 Discrete-time queueing network models for multiple access protocols 108
5.4 Equilibrium point analysis 114
5.5 Different customer classes 120
5.5.1 Recursive EPA 120
5.5.2 Extended EPA 125
5.6 Exercises 128

**6 SATELLITE NETWORKS** 129
6.1 Time-division multiple access 130
6.2 Slotted Aloha 134
6.2.1 Zero channel delay 134
6.2.2 Different customer classes 137
6.2.3 Finite channel delay 141
6.3 Code division multiple access 149
6.4 Buffered slotted Aloha 154
6.5 Exercises 160

**7 LOCAL AREA NETWORKS** 162
7.1 Carrier sensing networks 164
7.2 Token passing networks 171
7.2.1 Token bus and token ring 171
7.2.2 Timed token protocols 178
7.3 Slotted rings 186
7.4 Exercises 195

**APPENDIX** 197

**REFERENCES** 198

**INDEX** 202

# 1. NETWORKS, QUEUES AND PERFORMANCE MODELLING

*Chapter Objectives: To discuss the aims and philosophy of the book as a whole, describe the types of networks that are to be modelled, and introduce the concepts of discrete-time queues and their use as a modelling tool.*

## 1.1 INTRODUCTION

The following pages are about statistical prediction of the behaviour of communication networks for computers and other similar sources of digital information.

The typical sorts of predictions that we would like to make include:

- Which network protocol gives the best delay – throughput characteristic under specified conditions?
- What size buffers must be employed by a network's users to keep the probability of buffer overflow below a particular value?
- What is the maximum number of voice calls that can be accepted by a network in order to keep the voice packet transfer delay bounded?
- How many users can a satellite link support and still maintain a reasonable response time?

These and other related questions can be answered by developing a *performance model* for the network in question, and then solving this to obtain *performance measures.*

In this respect, most performance modelling techniques are based on continuous-time concepts, and these can often be difficult to apply to digitised computer communication networks. Such systems lend themselves in a natural way to a discrete-time approach, where the basic time unit can be used to reflect such things as time slotting in the physical system.

The following can therefore be described as an attempt to fill this gap, and present a unified, discrete-time approach for the modelling of computer communication networks and similar systems.

Before looking at specific application areas, a discrete-time theory for modelling and analysing the networks is first developed. This involves a gradual progression by starting with the necessary probability theory, then moving on through discrete-time Markov chains and discrete-time queues, to discrete-time queueing networks. These latter will be the main theoretical modelling tool that is used.

For applications, we concentrate principally on two basic types of networks: Wide area broadcast networks, such as satellite networks or ground-based radio networks, and local area networks. These latter, as their name implies, service a much smaller area than wide area networks. This area might consist typically of a building or college campus.

These two network types typify the modern trends in the subject area, and the modelling techniques used are generally applicable to other network types.

## 1.2 NETWORK TYPES

We shall consider a communication or computer network to consist of a population of $N$ geographically distributed *users* connected by a communications facility such as a channel (or a number of channels), arranged according to some *network topology*. In this context, the term 'user' denotes an access point to the communications facility. Located here might be, for example, a number of user-terminals, and information processing devices (such as computers) with which the terminals interact. These devices can be accessed by other users by establishing some form of connection via the network. In this way, the entire catchment of information processing devices is made available to all of the network's users. The general arrangement is shown in Fig. 1.1.

The types of network that will be considered are satellite networks and local area networks. As previously mentioned, these two types of networks typify the modern trends in the subject, and provide a sufficient application area to generalise the associated performance modelling techniques.

In the case of satellite networks, the topology is very simple in that we have $N$ geographically distributed users connected by a single channel which forms the only means of communication between them. All information thus has to be exchanged via the channel. The key feature of a satellite network, at least from a performance modelling point of view, is that the channel has a very large (in communication terms) propagation

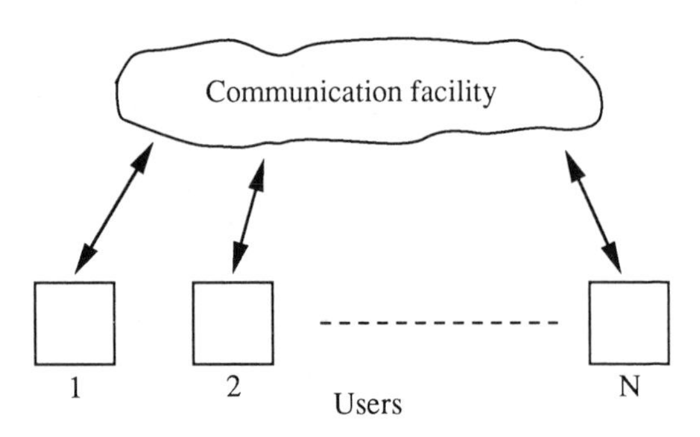


*Figure 1.1 Communication or computer network*

delay. This is very much in evidence in interactive situations, such as live television interviews via satellite. This delay has to be taken into account in modelling the network and, as will subsequently be seen, can lead to not inconsiderable complications in certain cases.

Local area networks are those that span a geographically small area, such as a building or college campus. We shall consider these to have either a *bus* or *ring* topology. In the case of a bus topology, the $N$ users are tapped at various points onto a length of coaxial cable or other similar bidirectional transmission medium. The system is thus again characterised by a single shared channel, but this time with a very small end-to-end propagation delay due to the small distances involved.

The other major local area network topology is the ring topology, where the $N$ users are connected via interfaces to a unidirectional loop of cable or optical fibre. Again delays are small, but very different operating principles are called for than for the bus topology.

Figure 1.2 summarizes the Topologies for the types of networks to be considered.

## 1.3 MULTIPLE ACCESS PROTOCOLS

The performance modelling of a network of one of the specified types is inherently concerned with the network's *multiple access protocol.* Multiple access refers to the capability to share a communication facility amongst a population of geographically distributed users. In what follows we shall always consider the communication facility that is to be shared to be the transmission medium (or channel). In this context the multiple access protocol is an algorithm for controlling access to the channel, and to efficiently allocate the channel capacity to the network's users.

In the case of local area networks, the multiple access protocol is also called the *medium access control protocol* [STAL 87], [HAMM 86]. This is usually abbreviated as *MAC protocol.*

A multiple access protocol can be identified as a protocol that overlaps both the Physical Layer and the Data Link Layer of the International Standards Organisation (ISO) Open System Interconnection (OSI) reference model [TANE 88]. In the case of local area networks, a MAC protocol is identified by its own specific layer in the IEEE 802 standards [STAL 87], [HAMM 86].

Figure 1.3 shows a classification of multiple access protocols. The *fixed assignment* methods dedicate a fixed portion of the available channel capacity to each user. The most common forms of this technique are *TDMA* (*Time Division Multiple Access*), *FDMA* (*Frequency Division Multiple Access*), and *CDMA* (*Code Division Multiple Access*).

In *random access* (contention) protocols, the entire bandwidth is presented to the users as a single channel to be accessed randomly. This means that *collisions* of messages can occur, and that colliding messages

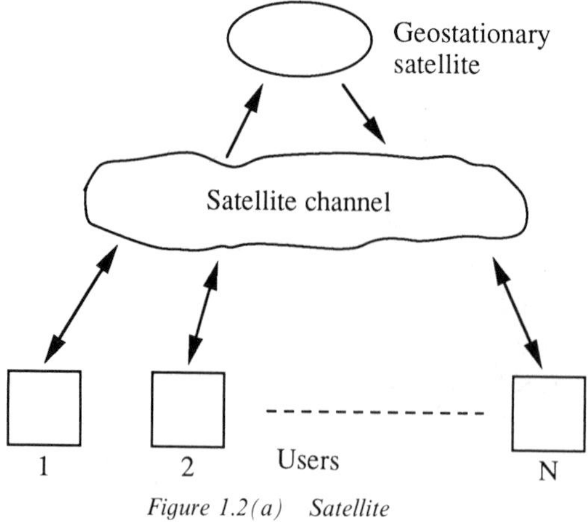


*Figure 1.2(a) Satellite*

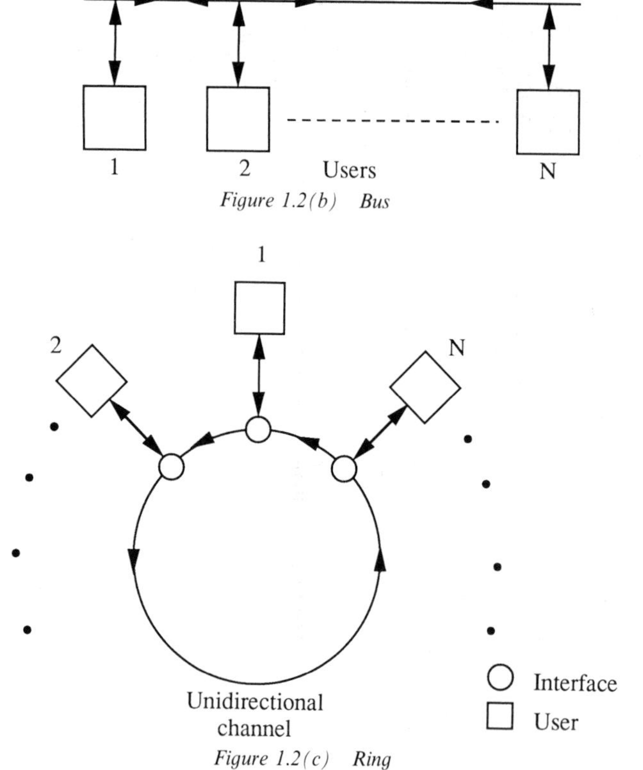


*Figure 1.2(b) Bus*

*Figure 1.2(c) Ring*

*Figure 1.2 Network topologies*

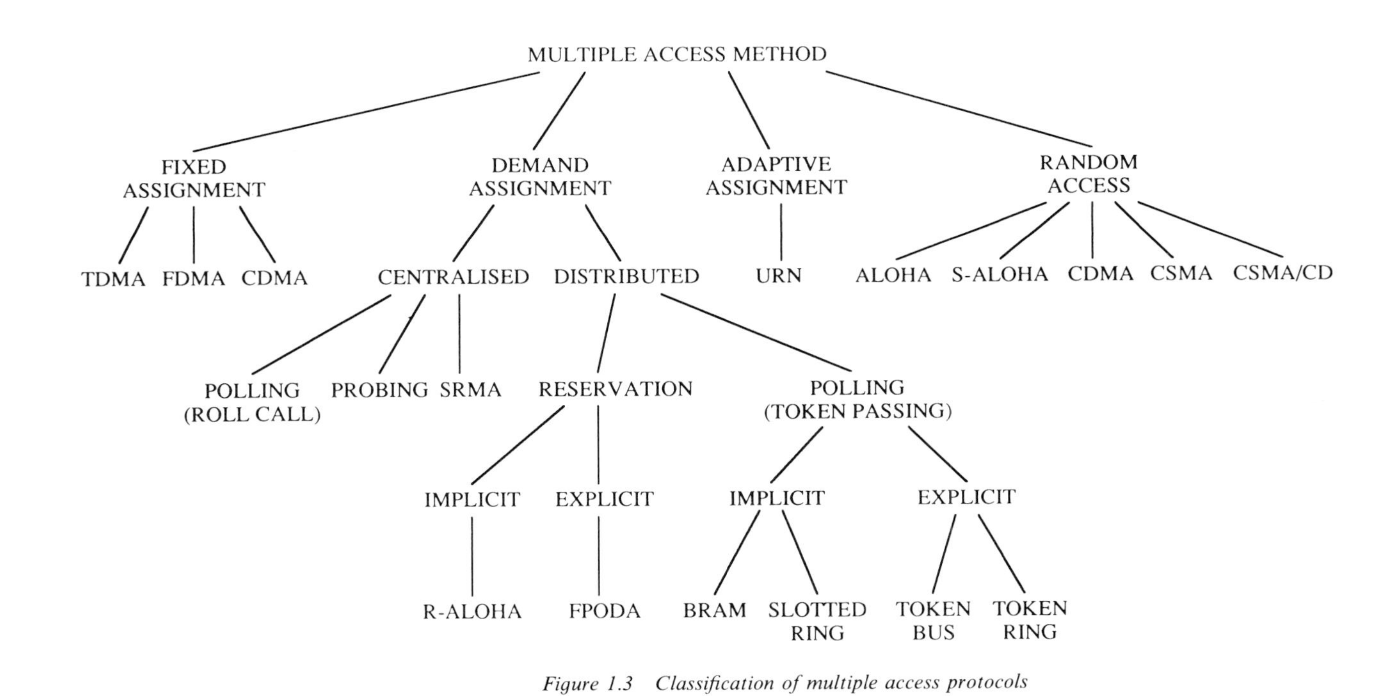


*Figure 1.3 Classification of multiple access protocols*

must be retransmitted. Random access protocols are mainly concerned with the retransmission scheme used and how the *contention* for the channel can be resolved. Examples of this type protocol are *S-ALOHA* (*Slotted ALOHA*) [ROBE 72], *CSMA* (*Carrier Sense Multiple Access*) and *CSMA/CD* (*Carrier Sense Multiple Access with Collision Detection*) [TOBA 80A].

*Demand assignment* methods require that explicit control information concerning the users' need for channel capacity is exchanged. The control can be either *centralised* or *distributed*; in a centralised scheme a central controller exists to decide which user should next have the channel access right, whereas with a distributed scheme, each user monitors the requests of all other users, and based on this information executes an identical distributed algorithm to determine who next has access to the channel. In both schemes the control information is exchanged through the channel, which implies additional overhead.

Examples of demand assignment with centralised control are *polling* [MART 72], *probing* [HAYE 78] and *SRMA* (*Split-Channel Reservation Multiple Access*) [TOBA 76B].

Examples of demand assignment with distributed control are the implicit reservation protocol *R-ALOHA* (*Reservation ALOHA*) [CROW 73], and the explicit reservation protocol based in TDMA called *FPODA* (*Fixed Priority Orientated Demand Assignment*) [JACO 78]. Other examples are implicit (virtual) token passing protocols such as the *BRAM* (*Broadcast Recognizing Access Method*) [CHLA 79] and the *slotted ring* [PIER 72], and the explicit token passing protocols such as the *token bus* [STAL 87], and the *token ring* [STAL 87].

The final class of multiple access protocols given in Fig. 1.3 are the *adaptive assignment* protocols. These have not been studied as extensively as the other schemes. Here the access method can change according to the loading offered by the users, the objective being to achieve near optimum performance at all times. The problem here is that the overhead incurred (sometimes a separate signalling channel is required) can often outweigh the advantages. A well known example of an adaptive scheme is the *URN* protocol [KLEI 78], [FUKU 81], so called because its analysis can be reduced to a familiar problem in probability theory in which coloured balls are withdrawn from an urn.

## 1.4 DISCRETE-TIME QUEUES

One of the major tools that will be used in modelling and analysing the performance of a network is the theory of *discrete-time queues*. The notation that is used to describe such systems is now introduced.

A queueing system can be described in terms of six components;

1. The *arrival process*: This is a stochastic process describing how

*customers* (customers are usually *messages* in a communications context) arrive into the system.
2. The *service process*: This is a stochastic process describing the length of time that a *server* (servers can be *communication channels* in a communications context) is occupied by a customer; for example, the length of time it takes to transmit a message in a computer communication network.
3. The *number of servers*.
4. The *waiting room*: The limit to the number of customers that can be accommodated to wait for service, including those currently being served.
5. The *customer population*: The limit to the total number of customers, including potential ones. This can be considered as part of the arrival process, since the number of potential customers can change the rate of arrivals.
6. The *service discipline*: The rules for deciding which customer or customers, within a queueing system, to serve.

The standard notation for specifying the first five of the above components is *Kendall's notation* [KEND 53], which consists of a series of letters and numbers separated into fields by a forward slash. In general, this notation can be represented as $A/B/c/n/p$, where $A$ and $B$ describe the arrival and service processes, respectively, $c$ gives the number of servers, $n$ the waiting room, and $p$ the customer population. The final two fields are optional, and if omitted are assumed, by default, to be infinite. Clearly, $c$, $n$ and $p$ can take on only positive integer values including infinity. The letters which describe the arrival and service processes are chosen from a small set of descriptors. The most common ones are:

$D$: This stands for *deterministic*, which implies constant interarrival or service times.

$M$: This stands for *Markovian* (or *memoryless*). In a continuous system, $A = M$ implies that *interarrival times are exponentially distributed*, so that the arrival process is *Poisson*. Similarly, service times are exponentially distributed if $B = M$. In a discrete-time system, $A = M$ implies that *interarrival times are geometrically distributed*. If $B = M$ then service times are also geometrically distributed.

$G$: This stands for *generally distributed*, which means that no restrictions are placed on the type of distribution that can be used.

Other descriptions are often used in discrete-time systems, such as *Geo* to represent geometrically distributed interarrival times. We shall retain the use of $M$ in this book however. Provided that a queue is explicitly described as either continuous or discrete-time, no ambiguity should arise.

A common extension of the above is to attach a superscript to a descriptor to indicate multiple (batch) arrivals. We shall use $a_n$ to denote batch size of the $n$th arrival, and $d_n$ to denote the batch size of the $n$th departure (service completion). In discrete-time systems $a_n$ denotes the number of arrivals in the $n$th time unit and $d_n$ the number of departures in the $n$th time unit. If $a_n$, $d_n \in \{0, 1\}$, $n = 0, 1, 2, \ldots$, then the superscript will be omitted.

To give some examples of the above notation, a continuous-time $M/M/1$ queue has a Poisson arrival process, exponentially distributed service times, a single server, infinite waiting room, and an infinite customer population. A discrete-time $M/M/1$ queue has geometrically distributed interarrival times with either zero or one arrival permitted in a time unit. This implies that the arrival process is a *Bernoulli process.* Also, this queue has a geometric service time distribution, a single server with either zero or one departure in a time unit, infinite waiting room, and an infinite customer population. A discrete-time $M^{a_n}/M/1$ queue is the same as a discrete-time $M/M/1$ queue except that the number of customers arriving in time unit $n$ is a sequence of independent and identically distributed (i.i.d.) random variables $\{a_n, n = 0, 1, 2, \ldots\}$, the distribution of which can be specified. Note that the interarrival times of the batches is still geometrically distributed.

The sixth component describing a queueing system, the service discipline, is not included in the Kendall notation. Typical service disciplines are first-come first-served (FCFS), last-come first-served (LCFS), and processor sharing (PS). The first two have their obvious meanings, while the PS discipline implies that in each time unit a unit of service is distributed equally amongst all the customers in the queue.

Unless otherwise implied, the service discipline will be taken to be FCFS, which perhaps needs some clarification in the context of a discrete-time system when multiple (batch) arrivals can occur in a time unit. The $a_n$ customers arriving in time unit $n$ are assumed to randomly order themselves (within the batch) in serial fashion, and are then queued behind the $a_{n-1}$ customers that arrived in the previous time unit. In this way, when the queueing discipline is FCFS all customers arriving in a given time unit begin service before any customer that arrives in the next time unit.

As a simple example of a discrete-time queue, consider a FCFS discrete-time $M/D/1$ system. This might, for example, be used to model a user's buffer in a computer communication network. According to our notation, the $M$ indicates that this queue has geometrically distributed interarrival times, with, at most, one customer permitted to arrive in a time unit. The $D$ indicates that customer service times are constant, with, at most, one customer permitted to depart in a time unit. Also, the queue has a single server, infinite waiting room, and a potentially infinite customer population.

Let a single new customer arrive in a time unit with probability $\alpha$, and no new customer arrive with probability $1 - \alpha$. Furthermore, let us assume that the (constant) service time of a customer is one time unit. If we define the state of the system as the number of queued customers left to be served, then if the queue starts off empty, the only possible states are 0 and 1. Then if $x_n$ describes the state of the queue during time unit $n = 0, 1, 2, \ldots$, we can formally describe the system's behaviour by the sequence of i.i.d. random variables $\{x_n = x,\ n = 0, 1, 2, \ldots\}$, with $x_0 \equiv 0$ the *initial state*, and $x \in \{0, 1\}$ the *state space*.

If the queue is in state 0 it is empty, and makes a transition to state 1 at the next time unit with probability $\alpha$. When the queue is in state 1, if no arrival occurs then the queue will make the transition to state 0 at the next time unit, and if an arrival occurs the queue will remain in state 1 at the next time unit, since the arriving customer will exactly balance the customer that is served. Note that in the above it is usual to assume that arrivals occur just after the beginning of a time unit, and that departures (service completions) take place just before the end of a time unit.

Even if the system has an initial state that is different from 0, say a long initial queue of customers, with probability 1 it will always eventually return to state 0, and once that state has been reached, the only other state that can be reached is state 1. In this case any state other than 0 or 1 is called *transient* in that the system can only assume such a state once, and can never return to that state once it has left it.

In most instances, we are interested in the long run performance of a system, rather than its transient performance. In other words, we are interested in the behaviour of a system when observed over a very long period of time, and, in particular, the probability of finding a system in a particular state when observed at a random time unit. Such probabilities are the *equilibrium probabilities*. The equilibrium probability of any transient states is clearly 0.

In the discrete-time $M/D/1$ queue, it is rather obvious that the equilibrium probabilities of states 0 and 1 are $1 - \alpha$ and $\alpha$ respectively.

There are, of course, formal ways of calculating equilibrium probabilities, or at least obtaining approximations for them. Such calculations are central to the theme of performance evaluation of systems such as computer communication networks, since most performance measures of interest can be derived directly from the equilibrium probabilities.

Also, it is worth noting that the behaviour of queueing systems can be conveniently represented by *state transition diagrams*. That for the discrete-time $M/D/1$ queue is shown in Fig. 1.4, where the nodes in the

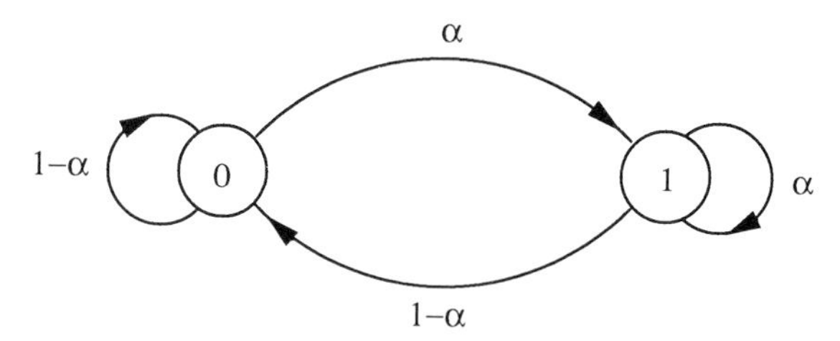


*Figure 1.4 State transition diagram for a discrete-time M/D/1 queue*

diagram represent states, and the directed arcs between nodes represent possible state transitions, and, as such, are labelled with the *transition probabilities.* This diagram has been derived on the basis of the state being observed just after the arrivals, this convention being somewhat flexible in a discrete-time system.

Had the state been observed at the time unit boundaries (just before the arrivals and just after the departures), which is another possible convention, then state 1 becomes transient in that once the system is in state 0 it can never leave that state. This can be seen in that if an arrival takes place just after the state has been observed as $x_n = 0$, then this customer must depart just before the state is observed at time $n + 1$, resulting in $x_{n+1} = 0$. This is a phenomenon that is unique to discrete-time systems, and illustrates the importance of having a sufficiently small time unit in relation to the interarrival and service times of customers, otherwise information concerning the system's behaviour can be lost. In the limit, when the time unit tends to zero, we have the familiar continuous-time systems.

## 1.5 PERFORMANCE MEASURES

There are many important measures that can be used for assessing the performance of a communication or computer network. Among these are throughput, average message delay, probability of message loss, reliability, and adaptability to variations in traffic.

The two main performance measures however are *throughput*, defined as the number of successfully transmitted messages per mean transmission time of a message, and the *message delay*, defined as the time interval, in units of the average transmission time of a message, from the moment a message is generated, to the instant it is correctly received. These will be adopted as the key performance measures for all the networks considered. It should be noted that it is sometimes more convenient to express these performance measures in units of *slots* rather than average message transmission times, where a slot is the basic, fixed time unit in terms of which all other times, such as message transmission times, channel propagation delays, and so on, are measured.

The mean values of throughput and message delay will be given the symbols $S$ and $W$ respectively.

A third performance measure that will often be used is the *probability of buffer overflow*, $B$. This is defined as the fraction of messages that are lost due to no buffer space being available at the time of their arrival. This measure can be important in assessing the transmission quality for certain types of data such as voice or video.

As a simple example of calculating these measures, we can consider the discrete-time $M/D/1$ queue explained in Section 1.4 as a model for a user's buffer in a computer communication network. The mean throughput for this user will then simply be the fraction of time the user spends transmitting messages. In other words, it will be the fraction of time that the queueing system spends serving customers. Noting that a customer will be served in a time unit (henceforth called a slot) only if the queue is in state 1, then if $\Pi_i$ is the equilibrium probability that the queue is in state $i$, clearly the mean throughput is

$$S = \Pi_1 = \alpha \tag{1.1}$$

This is in units of messages per slot, or messages per mean message transmission time, this latter being equal to one slot in this particular case.

The mean message delay can be calculated by first finding the average number of messages stored in the system, and then using the famous *Little's result* [LITT 61]. This result will be considered later, but simply states that the average number of customers stored in a queueing system, $L$, is equal to the product of the average rate of customer arrivals, $\lambda$, and the average time a customer spends in the system, $W$. This latter, of course, is the mean waiting time or delay through the system. We thus have

$$L = \lambda W \tag{1.2}$$

Clearly, for the discrete-time $M/D/1$ queue we must have

$$L = 0.\Pi_1 + 1.\Pi_1 = \alpha \tag{1.3}$$

and

$$\lambda = \alpha \tag{1.4}$$

This gives the rather trivial result that the mean message delay is equal to 1 slot, which is somewhat obvious from intuition. Also obvious is that the probability of buffer overvlow is zero for this system. This would be true, even if the buffer had a finite capacity.

# 2. PROBABILITY, RANDOM VARIABLES, AND DISTRIBUTIONS

*Chapter Objectives: To cover the basic probability theory required in the remainder of the text.*

The theory of probability is a mathematical tool that can be used to quantify uncertainty, and in the context of a communication or computer network it provides an essential means of analysing the behaviour of these systems. At first sight, it may appear odd that a seemingly deterministic entity needs to be treated in the same way as some random phenomena, but the number of variables that interact to influence a system's performance can be so large that the resulting behaviour is essentially unpredictable, and methods based on probability theory are often the only way forward.

It is anticipated that readers have had some prior exposure to this theory, and so the treatment here serves mainly as a review and reference source for subsequent chapters, and also to establish a notation for the latter; it is not intended to be comprehensive. Those readers with little previous exposure will benefit from using some suitable engineering level texts to fill in the details, such as those by Feller [FELL 71] or Papoulis [PAPO 84]. The treatment here follows [MITR 87] and [PAPO 84].

## 2.1 PROBABILITY

Let us first consider an *experiment*, such as rolling a die. The *sample space* of the experiment, usually denoted by the symbol $\Omega$, consists of the set of all possible outcomes. Thus, in the case of the die experiment

$$\Omega = \{1, 2, 3, 4, 5, 6\} \tag{2.1}$$

where the integers 1 through 6 represent the number of dots on the six faces of the die. These six outcomes are the *sample points* of the experiment. Note that, in general, the sample space may be a finite, denumerable, or non-denumerable set, and its membership depends on how we define the term 'outcome'.

An event is a subset of $\Omega$, which may consist of any number of sample points. Thus if we define an event $A$ as

$$A = \{2, 4\} \tag{2.2}$$

this event consists of outcomes 2 and 4. The *complement* of the event $A$, denoted by $A^c$, consists of all sample points in $\Omega$ that are not in $A$. Thus

$$A^c = \{1, 3, 5, 6\} \tag{2.3}$$

Two events are said to be *mutually exclusive* if they have no sample points in common. For example, if we define an event

$$B = \{1, 3, 6\} \tag{2.4}$$

then $A$ and $B$ are mutually exclusive. Clearly, by definition of complement, an event and its complement must always be mutually exclusive.

The *union* of two events is an event that consists of all the sample points in the two events. So if we define an event

$$C = \{1, 2, 3\} \tag{2.5}$$

then the union of $B$ and $C$, denoted by $B \cup C$, is the event

$$D = B \cup C = \{1, 2, 3, 6\} \tag{2.6}$$

Clearly, then, the union of an event and its complement form the entire sample space $\Omega$.

The *intersection* of two events is an event that consists of all the sample points that are common to both of the original events. The intersection of $B$ and $C$, denoted by $B \cap C$, or more simply by $BC$ where no ambiguity arises, is the event

$$E = BC = \{1, 3\} \tag{2.7}$$

There is considerable freedom in deciding which subsets of $\Omega$ are to be classified as events, and which are not. Whatever definition of an event is used it is necessary that all of the previously defined operations can be carried out. More precisely, the set of all events must satisfy the following three axioms:

Axiom 1. The entire sample space, $\Omega$, is an event (this event always occurs, not matter what the outcome of the experiment).

Axiom 2. If $A$ is an event, then $A^c$ is also an event.

Axiom 3. If $\{A_1, A_2, \ldots\}$ is a finite or denumerable set of events, then the union $A = \bigcup_i A_i$ is also an event.

From Axioms 1 and 2 it follows that the empty set, $\phi$, is an event (an event that can never occur). Also, from Axioms 2 and 3, if $\{B_1, B_2, \ldots\}$ is a finite or denumerable set of events, the intersection

$$B = \bigcap_i B_i = \left(\bigcup_i B_i^c\right)^c$$

is also an event.

Having defined the events that may occur as a result of an experiment, we now require to formulate some means of measuring the *relative likelihoods* of these occurrences. We do this by assigning to each event, $A$, a number denoted by $P(A)$, called the probability of that event. The probabilities of all events, $A$, are such that $0 \leq P(A) \leq 1$. That is, the probabilities are normalised such that the probability of an event that is certain to occur is 1, and that of an event which cannot occur is 0.

In a similar way that we defined a set of axioms for events, the same can be done for the probabilities of those events. We define probability to be a function, $P(.)$, valid over the set of all events, such that the values of the function are real numbers, and the function satisfies the following three axioms:

Axiom 4. $0 \leq P(A) \leq 1$ for all possible events, $A$.
Axiom 5. $P(\Omega) = 1$.
Axiom 6. If $\{A_1, A_2, \ldots\}$ is a finite or denumerable set of disjoint (non-overlapping) events, then $P\left(\bigcup_i A_i\right) = \sum_i P(A_i)$

Note that Axiom 6 states that the probability of the union of a set of disjoint events is equal to the sum of the probabilities of the individual events, and is essential to give the probability function the additive property of measures.

A direct consequence of Axioms 5 and 6 is that if $\{A_1, A_2, \ldots\}$ is a partition of $\Omega$ (all the events, $A_i$, are disjoint and their union is $\Omega$), then

$$\sum_i P(A_i) = 1 \tag{2.8}$$

This means that for every event $A$,

$$P(A^c) = 1 - P(A) \tag{2.9}$$

Hence the probability of the empty event is

$$P(\phi) = 1 - P(\Omega) = 0 \tag{2.10}$$

Interestingly, the probability of an event being 0 does not necessarily mean that it can never occur. To consider the previous die rolling experiment, if we construct a new experiment that a fair die is rolled an infinite number of times, it can be shown that the probability

$$P(\text{a 6 never appears}) = 0$$

yet the event can occur.

To construct a probability function when the sample space is finite or denumerable, it suffices to assign the $i$th outcome of an experiment a non-negative weight, $p_i$, so that

$$\sum_i p_i = 1 \tag{2.11}$$

where

$$p_i = p(A_i) \tag{2.12}$$

Thus if we roll a fair die, then

$$p_1 = p_2 = \cdots = p_6 = \tfrac{1}{6} \tag{2.13}$$

If $A$ and $B$ are two arbitrary events then, as a consequence of the set identity

$$A \cup B = A \cup (A^c B)$$

and the fact that $A$ and $A^c$ are disjoint events

$$P(A \cup B) = P(A) + P(A^c B) \tag{2.14}$$

Likewise, since we have the set identity

$$B = (AB) \cup (A^c B)$$

then

$$P(B) = P(AB) + P(A^c B) \tag{2.15}$$

From equations (2.14) and (2.15) we have

$$P(A \cup B) = P(A) + P(B) - P(AB) \tag{2.16}$$

This result can be generalised in that if $A_1, A_2, \ldots$ are arbitrary events, then

$$\begin{aligned} P\left(\bigcup_i A_i\right) &= P(A_1) + P(A_1^c A_2) + P(A_1^c A_2^c A_3) + \cdots \\ &\leq \sum_i P(A_i) \end{aligned} \tag{2.17}$$

The equality in (2.17) only holds if $A_1, A_2, \ldots$ are disjoint. Thus the union of a set of arbitrary events is only equal to the sum of their individual probabilities if those events are disjoint.

Equivalently, the probability of the intersection of two events is not necessarily equal to the product of their individual probabilities. This is true only if the outcome of one event does not influence the outcome of the other, in which case the events are said to be *independent*. Conversely, two events $A$ and $B$ are independent if

$$P(AB) = P(A)P(B) \tag{2.18}$$

For example if a fair die is rolled twice, and events $A$ and $B$ represent the even numbered possible outcomes as a result of the first and second rolls respectively, then

$$A = B = \{2, 4, 6\} \tag{2.19}$$

Thus

$$P(A) = P(B) = \tfrac{1}{2} \tag{2.20}$$

Now the event

$AB =$ (Even numbered outcome on first roll and even numbered outcome on second roll

is the probability of the nine pairs of outcomes $(i, j)$, $i = 2, 4, 6$, $j = 2, 4, 6$, which is 1/4.

Thus

$$P(AB) = \tfrac{1}{4} \tag{2.21}$$

and from (2.20),

$$P(A)P(B) = \tfrac{1}{4} \tag{2.22}$$

so the events $A$ and $B$ are independent.

The definition of independence for two events can be extended to an arbitrary, finite set of $n$ events $\{A_1, A_2, \ldots, A_n\}$, $n \geq 3$, in that these $n$ events are said to be *mutually independent* if

(i) $P(A_1, A_2, \ldots, A_n) = P(A_1)P(A_2)\ldots P(A_n)$, and
(ii) every subset of $n - 1$ events among them are mutually independent.

Thus three events $\{A_1, A_2, A_3\}$ are mutually independent if

$$\begin{aligned} P(A_1 A_2) &= P(A_1)P(A_2) \\ P(A_1 A_3) &= P(A_1)P(A_3) \\ P(A_2 A_3) &= P(A_2)P(A_3) \\ P(A_1 A_2 A_3) &= P(A_1)P(A_2)P(A_3) \end{aligned}$$

Note that neither (i) nor (ii) alone is a sufficient condition for independence.

The notion of independence is closely related to that of conditional probability. The conditional probability that an event $A$ occurs, given that an event $B$ has occurred, which is denoted by $P(A|B)$, is defined as

$$P(A|B) = \frac{P(AB)}{P(B)} \tag{2.23}$$

Clearly, if $A$ and $B$ are independent then

$$P(A|B) = P(A) \tag{2.24}$$

If we now consider the conditional probability of event $B$, given that $A$ has occurred, then the equivalent of equation (2.23) is

$$P(B|A) = \frac{P(AB)}{P(A)} \tag{2.25}$$

Then from (2.23) and (2.25) we have

$$P(AB) = P(A|B)P(B) = P(B|A)P(A) \tag{2.26}$$

Generalising this to more than two events then

$$P(A_1 A_2 \dots A_n) = P(A_1|A_2 \dots A_n)P(A_2|A_3 \dots A_n) \dots P(A_{n-1}|A_n)P(A_n) \tag{2.27}$$

It is often convenient to determine the probability of a given event, $A$, by *conditioning* it on the occurrence of one of several other events. Let $\{B_1, B_2, \dots\}$ be a partition of $\Omega$; then any event $A$ can be represented as

$$A = A\Omega = A\left(\bigcup_i B_i\right) = \bigcup_i AB_i \tag{2.28}$$

Since the events $AB_i$, $i = 1, 2, \dots$, are disjoint, then

$$P(A) = \sum_i P(AB_i) = \sum_i P(A|B_i)P(B_i) \tag{2.29}$$

This is called the *complete probability formula*, and enables us to find the probability of $A$ assuming the $P(B_i)$ are known and the $P(A|B_i)$ can be obtained.

This leads to the well known *Bayes' formula* that gives the conditional probability of some event $B_i$, given that the event $A$ has occurred. Using equation (2.25) we have

$$P(B_i|A) = \frac{P(B_i A)}{P(A)} \tag{2.30}$$

and from equation (2.26)

$$P(B_i|A) = \frac{P(A|B_i)P(B_i)}{P(A)} \tag{2.31}$$

Now using equation (2.29) for $P(A)$ in (2.31)

$$P(B_i|A) = \frac{P(A|B_i)P(B_i)}{\sum_j P(A|B_j)P(B_j)} \tag{2.32}$$

This is the result known as Bayes' formula.

To illustrate the application of both the complete probability formula and also Bayes' formula, consider the following example.

*Example 2.1: Complete probability and Bayes' formula*

We have four urns (probability theorists tend to use urns rather than other, more conventional utensils), each containing a number of red or green balls according to the distribution shown in Table 2.1. The idea is

**Table 2.1** DISTRIBUTION OF COLOURED BALLS IN EXAMPLE 2.1

| *Urn number* | *Number of green balls* | *Number of red balls* |
|---|---|---|
| 1 | 19 | 1 |
| 2 | 3 | 2 |
| 3 | 9 | 1 |
| 4 | 9 | 1 |

to *randomly* select *one* of the urns, and then *randomly* select from the chosen urn *one* ball.

(i) What is the probability that the selected ball is red?
(ii) Given that the selected ball is red, what is the probability that it came from urn 2?

Solution:

(i) Let $A$ be the event that the selected ball is red, and $B_i$, $i = 1, 2, 3, 4$, be the event that urn $i$ is selected.

Clearly,

$$P(B_1) = P(B_2) = P(B_3) = P(B_4) = \tfrac{1}{4} \tag{2.33}$$

Also,

$$P(A|B_1) = \tfrac{1}{20} = 0.05 \tag{2.34}$$

$$P(A|B_2) = \tfrac{2}{5} = 0.4 \tag{2.35}$$

$$P(A|B_3) = P(A|B_4) = \tfrac{1}{10} = 0.1 \tag{2.36}$$

Since $\{B_1, B_2, B_3, B_4\}$ form a partition of the complete sample space, $\Omega$, then using equation (2.29)

$$P(A) = \sum_{i=1}^{4} P(A|B_i)P(B_i) \tag{2.37}$$

Thus equations (2.33) through (2.37) give the result

$$P(A) = 0.1625 \tag{2.38}$$

(ii) To consider now the conditional probability that the selected ball came from urn 2, given that it is red, we require $P(B_2|A)$, which is given in the form of Bayes' formula by equation (2.32).

Then

$$P(B_2|A) = \frac{P(A|B_2)P(B_2)}{\sum_{i=1}^{4} P(A|B_i)P(B_i)} \tag{2.39}$$

$$= \frac{P(A|B_2)P(B_2)}{P(A)} \tag{2.40}$$

where the last equality came from (2.37). Thus using equations (2.33), (2.35) and (2.38) in (2.40) we have

$$P(B_2|A) = 0.615 \tag{2.41}$$

Note that the terms 'a priori' and 'a posteriori' are usually associated with probabilities such as the $P(B_i)$ and $P(B_i|A)$, respectively. Thus we say that the a priori probability of selecting urn 2 is 0.25, and the a posteriori probability that urn 2 was selected, given that the chosen ball is red is 0.615. In this case the partial information available, that the chosen ball is red, has enabled us to assign a higher probability weighting to the assertion that urn 2 was selected.

## 2.2 RANDOM VARIABLES

It is often required to consider the behaviour of functions defined on a sample space $\Omega$ and whose values are real numbers. These functions are called *random variables*, where the term 'random' here refers to the fact that the value of the function is known only after the experiment has been performed. The outcome of the experiment is called a *realisation* of the associated random variable.

The probability that a random variable, $X$, takes a value which does not exceed a given number, $x$, is called the *cumulative distribution function* (cdf) of $X$. This is a function of $x$, and is usually denoted by $F(x)$. Thus we can write

$$F(x) = P(X \leq x); \qquad -\infty < x < \infty \tag{2.42}$$

The cdf of any random variable has the following properties:

1. If $x \leq y$ then $F(x) \leq F(y)$.
2. $F(-\infty) = 0$ and $F(\infty) = 1$.
3. $F(x)$ is right continuous. That is, for any $x$ and any decreasing sequence $x_i$, $i \geq 1$, that converges to $x$, then $\lim_{i\to\infty} F(x_i) = F(x)$

If $a$ and $b$ are two real numbers such that $a < b$, then the probability that $X$ takes a value in the interval $(a, b]$ (where the notation $(a, b]$ denotes that the end point $A$ is not included in the interval, but the end point $b$ is) is given by

$$P(a < x \leq b) = F(b) - F(a) \tag{2.43}$$

If we let $a \to b$ in equation (2.43) then

$$P(x = b) = F(b) - F(b^-) \tag{2.44}$$

where $F(b^-)$ is the limit of $F(x)$ from the left, at point $b$.

We see immediately from equation (2.43) that if $F(a) = F(b)$ then

$$P(a < X \leq b) = 0 \tag{2.45}$$

and from equation (2.44), if $F(x)$ is continuous at point $b$ then

$$P(X = b) = 0 \tag{2.46}$$

Recall that an event can have zero probability, even though that event may occur.

Equations (2.45) and (2.46) lead us to distinguish between *continuous* random variables, and *discrete* random variables. For a continuous random variable, $X$, then $F(x)$ is a continuous function of $x$, and might have the form shown in Fig. 2.1(a). For any specified value of $x$, then according to equation (2.46) the probability that $X$ is equal to this value is zero. Thus for a continuous random variable, $X$, we can only define a non-zero probability for $X$ falling in some specific interval, say between $a$ and $b$, on the $x$-axis.

In the case of a continuous random variable, the derivative of the cdf, which is called the *probability density function* (pdf), plays an important role in calculating probabilities. The pdf is usually denoted by $f(x)$, and so we have

$$f(x) = \frac{dF(x)}{dx} \tag{2.47}$$

According to the definition of derivative, we can write

$$\begin{aligned} f(x) &= \lim_{\Delta x \to 0} \frac{F(x + \Delta x) - F(x)}{\Delta x} \\ &= \lim_{\Delta x \to 0} \frac{P(x < X \leq x + \Delta x)}{\Delta x} \end{aligned} \tag{2.48}$$

Then, when $\Delta x$ is small, $f(x)\Delta x$ gives an approximation to the probability that the random variable, $X$, takes a value in the interval $[x, x + \Delta x]$.

Clearly, from equation (2.47), given a pdf $f(x)$, the corresponding cdf is obtained by the integral

$$F(x) = \int_{-\infty}^{x} f(u)\, du \tag{2.49}$$

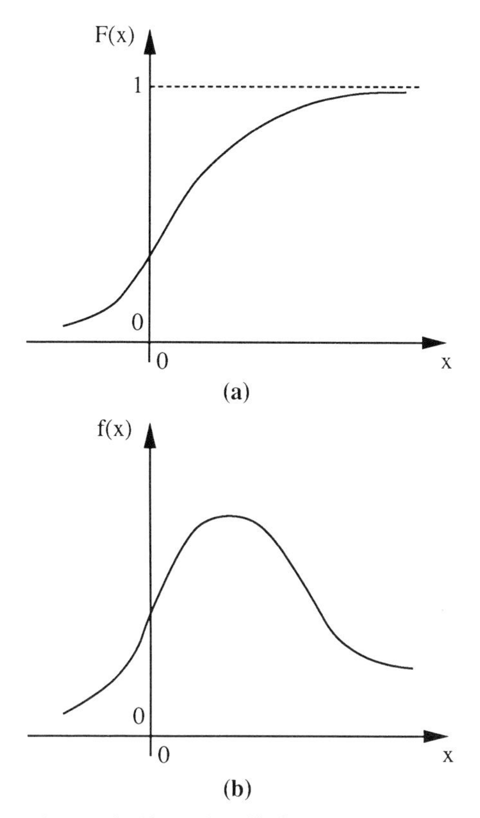


*Figure 2.1 (a) Typical cdf and (b) pdf of a continuous random variable X*

where, in order to ensure that $F(\infty) = 1$, we must have

$$\int_{-\infty}^{\infty} f(x)\,dx = 1 \tag{2.50}$$

We can thus use equation (2.49) to find the probability that $X$ falls in the interval $[a, b]$ as

$$P(a < x < b) = \int_{a}^{b} f(x)\,dx \tag{2.51}$$

Note that for a continuous random variable, it does not matter whether or not we include the end points of the interval.

A typical pdf curve for a continuous random variable is shown in Fig. 2.1(b). Note also that the probability that the random variable falls within a specified interval on the $x$ axis, as given by equation (2.51), is equal to the area under the pdf curve over this interval.

To return now to equation (2.44), if the function $F(x)$ is discontinuous at point $b$, then $F(b) \neq F(b^-)$ and so we now have

$$P(X = b) \neq 0 \tag{2.52}$$

If $F(x)$ is constant everywhere, except at a finite or denumerable set of points $x_1, x_2, \ldots$, where discontinuities occur with magnitudes $p_1, p_2, \ldots$, respectively, such that $p_1 + p_2 + \cdots = 1$, then the cdf curve has the form of a staircase function, as shown in Fig. 2.2(a). A random variable, $X$, that has such a cdf is called 'discrete', and assumes the values $x_i$ with probabilities $p_i$. Thus we have

$$P(X = x_i) = p_i \tag{2.53}$$

In the case of a discrete random variable, the pdf is called the *probability mass function* (pmf) and is denoted by

$$f(x) = \sum_i p_i \delta(x - x_i)$$

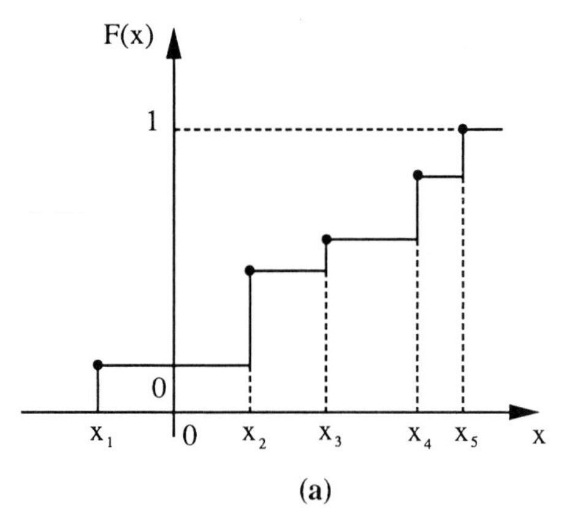


(a)

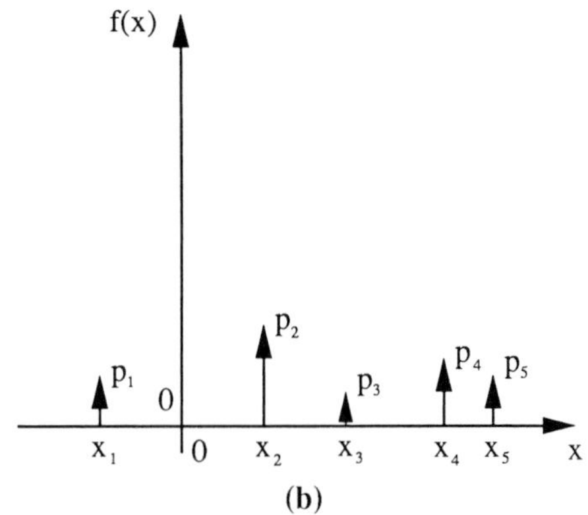


(b)

*Figure 2.2 (a) Typical cdf and (b) pdf of a discrete random variable X*

where $\delta(x - x_i)$ is a unit impulse located at $x = x_i$. A typical pmf is shown in Fig. 2.2(b), where the term $p_i\delta(x - x_i)$ is shown as a vertical arrow at point $x = x_i$ with length equal to $p_i$.

*Example 2.2*

Consider again the experiment of rolling a fair die, and define a random variable $X = 10i$, where $i$ is the number of dots on each face of the die. In this case we have

$$x_1 = 10, x_2 = 20, \ldots, x_6 = 60$$

and

$$p_1 = p_2 = \cdots = p_6 = 1/6$$

Then

$$f(x) = \tfrac{1}{6}[\delta(x - 10) + \delta(x - 20) + \cdots + \delta(x - 60)]$$

and so the cdf and pmf have the forms shown in Fig. 2.3(a) and Fig. 2.3(b), respectively.

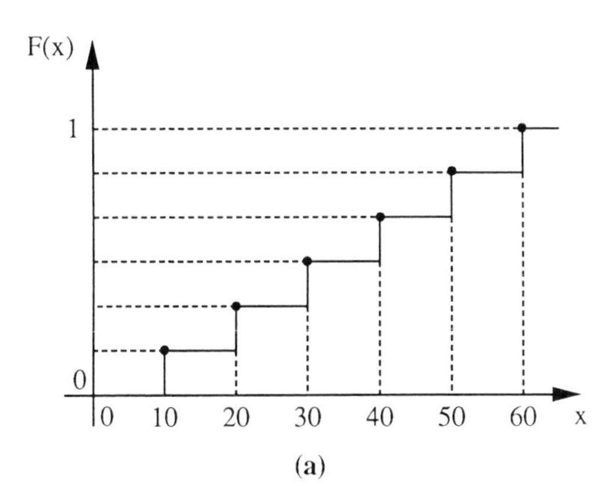


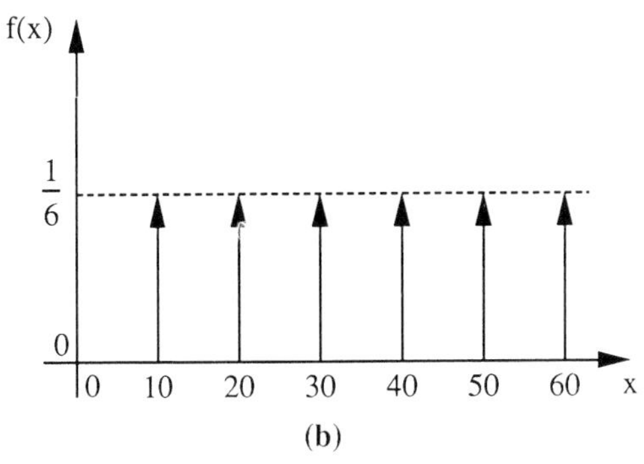


*Figure 2.3 (a) cdf and (b) pmf for Example 2.2*

To consider some specific cases, note that,

$$\begin{aligned}
F(100) &= P(X \leq 100) = P(\Omega) = 1 \\
F(30.01) &= P(X \leq 30.01) \\
&= p_1 + p_2 + p_3 \\
&= \tfrac{1}{2} \\
F(30) &= P(X \leq 30) \\
&= p_1 + p_2 + p_3 \\
&= \tfrac{1}{2} \\
F(29.99) &= P(X \leq 29.99) \\
&= p_1 + p_2 \\
&= \tfrac{1}{3}
\end{aligned}$$

### 2.2.1 Moments of a random variable

Although a random variable is characterised completely by either its cdf or pdf/pmf, such a characterisation is not always possible to obtain. Luckily, it is not always necessary to obtain a full characterisation in that it is usually sufficient to describe a random variable by a set of numbers, known as *moments*, that summarise the essential attributes of the random variable. These moments are defined in terms of the cdf, but can usually be determined directly without a knowledge of that function.

The *first moment*, which is more commonly called either the *mean* or the *expectation*, of a random variable $X$ is denoted by $E[X]$, and defined as the Riemann-Stieltjes integral

$$E[X] = \int_{-\infty}^{\infty} x dF(x) \tag{2.54}$$

If $X$ is a continuous random variable with pdf given by $f(x)$, then equation (2.54) becomes

$$E[X] = \int_{-\infty}^{\infty} x f(x)\, dx \tag{2.55}$$

For a corresponding discrete random variable $X$, taking values $x_1, x_2, \ldots$, with probabilities $p_1, p_2, \ldots$, respectively, then the integral in (2.55) becomes a sum

$$E[X] = \sum_i x_i p_i \tag{2.56}$$

In general we define the $n$th *moment* of a continuous random variable,

$X$, as

$$E[X^n] = \int_{-\infty}^{\infty} x^n f(x)\,dx \tag{2.57}$$

and the $n$th moment of a discrete random variable as

$$E[X^n] = \sum_i x_i^n p_i \tag{2.58}$$

By far the most important moments of $X$ are the first two moments. Thus putting $n = 1$ in either (2.57) or (2.58) gives the mean, and putting $n = 2$ gives the second moment, or *mean-square* value of $X$. Thus we have

$$E[X^2] = \int_{-\infty}^{\infty} x^2 f(x)\,dx \tag{2.59}$$

as the mean-square value of a continuous random variable, and

$$E[X^2] = \sum_i x_i^2 p_i \tag{2.60}$$

as the mean-square value of a discrete random variable $X$.

We may also define *central moments*, which are simply the moments of the difference between a random variable $X$ and its mean value $E[X]$. Thus the $n$th central moment is

$$E[(X - E[X])^n] = \int_{-\infty}^{\infty} (x - E[X])^n f(x)\,dx \tag{2.61}$$

for a continuous random variable $X$, and is

$$E[(X - E[X])^n] = \sum_i (x_i - E[X])^n p_i \tag{2.62}$$

for a corresponding discrete random variable.

The second central moment is extremely important in characterising a random variable in that it gives a measure of the 'spread' of a random variable around its mean. This is called the *variance* of $X$, and is defined in the continuous case as

$$\begin{aligned} \mathrm{Var}[X] &= E[(X - E[X])^2] \\ &= \int_{-\infty}^{\infty} (x - E[X])^2 f(x)\,dx \end{aligned} \tag{2.63}$$

and in the discrete case as

$$\begin{aligned} \mathrm{Var}[X] &= E[(X - E[X])^2] \\ &= \sum_i (x_i - E[X])^2 p_i \end{aligned} \tag{2.64}$$

It is common to give the variance of a random variable $X$ a special symbol, $\sigma_X^2$. The square root of the variance, namely, $\sigma_X$, is called the *standard deviation* of the random variable $X$.

By specifying the variance, $\sigma_X^2$, we essentially define the effective width of the pdf or pmf of the random variable $X$ about the mean $E[X]$. A precise statement of this definition is due to Chebyshev. The *Chebyshev inequality* states that for any positive number $\varepsilon$, we have

$$P(|X - E[X]| \geq \varepsilon) \leq \frac{\sigma_X^2}{\varepsilon^2} \tag{2.65}$$

From this inequality we see that the mean and variance of a random variable give a partial description of its cdf.

The expectation operator, $E[.]$, has a number of useful properties which we state without proof in the following.

1. If $X \geq 0$, then $E[X] \geq 0$.
2. If $c$ is a constant, then $E[cX] = cE[X]$.
3. If $X$ and $Y$ are two random variables, then

   $$E[X + Y] = E[X] + E[Y]$$

   Note that this property holds regardless of whether $X$ and $Y$ are independent or not.
4. $E[1] = 1$
5. If $X_1, X_2, \ldots$ is an increasing or decreasing sequence of random variables converging to $X$, then

   $$E[X] = \lim_{n \to \infty} E[X_n]$$

   This is a consequence of the fact that limit and integration can be interchanged.

From property 1 it follows that inequalities between random variables carry through to their expectations. That is,

$$\text{if } X \leq Y \text{ then } E[X] \leq E[Y] \tag{2.66}$$

The reverse of this consequence does not, however, necessarily hold.

The properties 2 and 3 can be combined into a more general one in that if $X_1, X_2, \ldots, X_n$ are arbitrary random variables and $c_1, c_2, \ldots, c_n$ are constants, then

$$E\left[\sum_{i=1}^{n} c_i X_i\right] = \sum_{i=1}^{n} c_i E[X_i] \tag{2.67}$$

The relation between the second moment, $E[X^2]$, and the second central moment, $E[(X - E[X])^2]$, is of interest in its own right, and can be obtained by expanding the bracket of the second central moment as follows.

$$\begin{aligned}\operatorname{Var}[X] &= E[(X - E[X])^2] \\ &= E[X^2 - 2XE[X] + (E[X])^2] \\ &= E[X^2] - 2E[X]E[X] + (E[X])^2 \\ &= E[X^2] - (E[X])^2 \end{aligned} \tag{2.68}$$

Recalling that the second moment, $E[X^2]$, is the mean-square value of $X$, we see that the variance, or second central moment, of a random variable $X$, is equal to the mean-square value of $X$ minus the square of the mean value of $X$.

Using equation (2.68) we can obtain an expression for the variance of the sum of two random variables, $X$ and $Y$:

$$\begin{aligned}\operatorname{Var}[X + Y] &= E[(X + Y)^2] - (E[X] + E[Y])^2 \\ &= E[X^2] + 2E[XY] + E[Y^2] - (E[X])^2 \\ &\quad - 2E[X]E[Y] - (E[Y])^2 \end{aligned} \tag{2.69}$$

Thus

$$\operatorname{Var}[X + Y] = \operatorname{Var}[X] + \operatorname{Var}[Y] + 2(E[XY] - E[X]E[Y]) \tag{2.70}$$

Clearly the variance of a sum of random variables $X$ and $Y$ is only equal to the sum of their variances if the quantity $E[XY] - E[X]E[Y]$ vanishes. This quantity is called the *covariance* of $X$ and $Y$, and is denoted by $\operatorname{Cov}[X, Y]$. Thus we have

$$\operatorname{Cov}[X, Y] = E[XY] - E[X]E[Y] \tag{2.71}$$

The covariance is a measure of the extent to which the corresponding random variables are *correlated*. If $\operatorname{Cov}[X, Y] = 0$ then $X$ and $Y$ are said to be *uncorrelated*. Clearly, if $X$ and $Y$ are independent, then since $E[XY] = E[X]E[Y]$, they must be uncorrelated. The converse of this, however, is not necessarily true.

It is also interesting to note that if $X + Y$ is replaced by $X - Y$ in (2.69), then the result is

$$\operatorname{Var}[X - Y] = \operatorname{Var}[X] + \operatorname{Var}[Y] - 2(E[XY] - E[X]E[Y]) \tag{2.72}$$

Comparing this expression with (2.70) we see that the variance of the sum of two random variables is also equal to the variance of the difference of the two random variables provided that they are uncorrelated.

### 2.2.2 Generating functions

There are many instances in practice where the quantity of interest is a sum of independent random variables. The mean and variance of such a sum can be obtained as shown in the previous subsection by adding together the means and variances, respectively, of the constituent variables.

To find the cdf or pdf (or pmf) of a sum of random variables is, however, a more difficult task. In subsequent chapters, we shall, for the most part, be concerned with discrete random variables; let us therefore consider the sum, $S$, of two discrete random variables $X$ and $Y$ that are independent, and whose values are non-negative integers. Denote by $p_i$, $q_i$ and $r_i$ $(i = 0, 1, 2, \ldots)$ the distributions of $X$, $Y$ and $S$, respectively. That is:

$$p_i = P(X = i); \qquad q_i = P(Y = i); \qquad r_i = P(S = i); \qquad i = 0, 1, 2, \ldots \tag{2.73}$$

Since for $S$ to take a value $i$, $X$ must take a value from $0, 1, \ldots, i$, and $Y$ must take a value $i - X$, we express $r_i$ as

$$\begin{aligned} r_i &= P(X + Y = i) \\ &= \sum_{j=0}^{i} P(X = j, Y = i - j) \\ &= \sum_{j=0}^{i} P(X = j)P(Y = i - j) \\ &= \sum_{j=0}^{i} p_j q_{i-j} \qquad i = 0, 1, \ldots \end{aligned} \tag{2.74}$$

Note that the third equality came from the fact that the variables are independent. The distribution of $r_i$, as given by (2.74), is called the *convolution* of the distributions $p_i$ and $q_i$. The convolution operation is often written, for short, using an asterisk; thus

$$r_i = p_j * q_{i-j} \qquad \begin{array}{l} i = 0, 1, \ldots \\ j = 0, 1, \ldots, i \end{array}$$

The convolution operation has, however, an alternative form which is often much more convenient to use, and which we shall next proceed to derive. First associate with the function $p_i$ the function $p(z)$, defined as

$$p(z) = \sum_{i=0}^{\infty} p_i z^i \tag{2.75}$$

This is called the *generating function*, or alternatively, the *z-transform* of the probabilities $p_i$. Since from equation (2.11)

$$\sum_i p_i = 1 \tag{2.76}$$

then we must have

$$p(1) = 1 \tag{2.77}$$

Thus, $p(z)$ must be finite for all real values of $z$ such that $|z| \leq 1$. The generating function, $p(z)$, has the important property that if $p(z)$ is known, then the probabilities $p_i$ are uniquely determined as

$$p_i = \frac{1}{i!}p^{(i)}(0) \qquad i = 0, 1, \ldots \tag{2.78}$$

where $p^{(i)}(0)$ is the $i$th derivative of $p(z)$ at $z = 0$. This, in effect, justifies the name 'generating function' in the sense that, from it, the corresponding probabilities can be generated.

It is rather obvious by comparing equations (2.56) and (2.75) that the generating function can also be expressed in the form of an expectation:

$$p(z) = E[z^X] \tag{2.79}$$

where the random variable $z^X$ takes a value $z^i$ with probability $p_i$, $i = 0, 1, \ldots$ .

From equation (2.79) we see that, given a generating function, it is quite easy to obtain the mean, and, indeed, the higher moments of $X$. This can be done by differentiating (2.79) with respect to $z$ and evaluating the result at $z = 1$. Thus, the mean of $X$ is obtained by

$$p^{(1)}(z) = E[Xz^{X-1}] \tag{2.80}$$

and so

$$p^{(1)}(1) = E[X] \tag{2.81}$$

The second moment can be obtained from the second derivative of $p(z)$ evaluated at $z = 1$. Thus

$$p^{(2)}(z) = E[X(X-1)z^{X-2}] \tag{2.82}$$

and setting $z = 1$ gives

$$\begin{aligned} p^{(2)}(1) &= E[X(X-1)] \\ &= E[X^2] - E[X] \end{aligned} \tag{2.83}$$

From this we can find the second moment, $E[X^2]$. In general

$$p^{(n)}(1) = E[X(X-1)\ldots(X-n+1)] \tag{2.84}$$

The right hand side of (2.84) is called the *factorial moment* of $X$, of order $n$. By expanding this, we can thus find the $n$th moment of a random variable $X$, for which reason the name *moment generating function* is sometimes used to describe $p(z)$.

From our point of view, however, the most important use of a generating function is that it provides us with a convenient method of finding the distribution of a sum of random variables. For example, if we wish to calculate the distribution of the random variable $S = X + Y$, where, recall that $X$ and $Y$ are discrete random variables that are independent, then introduce the generating functions $q(z)$ and $r(z)$ of the distributions $q_i$ and $r_i$, respectively.

Then, in a similar way to equation (2.79), we have

$$q(z) = E[z^Y] \tag{2.85}$$

and

$$r(z) = E[z^S] \tag{2.86}$$

Thus

$$\begin{aligned} r(z) &= E[z^{X+Y}] \\ &= E[z^X z^Y] \\ &= E[z^X]E[z^Y] \\ &= p(z)q(z) \end{aligned} \tag{2.87}$$

Thus we have replaced the convolution operation of equation (2.74) with a multiplication. In other words, the generating function of the convolution of two discrete distributions is equal to the product of their generating functions. In general, if $S$ is a sum of $n$ independent random variables, $X_1, X_2, \ldots, X_n$, that have generating functions $x_1(z), x_2(z), \ldots, x_n(z)$, respectively, then the generating function of $S$, call this $r(z)$, is given by

$$r(z) = x_1(z)x_2(z)\ldots x_n(z) \tag{2.88}$$

Thus the problem of finding the distribution of a sum of discrete, independent random variables reduces to finding the generating functions of the individual random variables, taking the product of these to find $r(z)$, and then using equation (2.78) to find the required probabilities.

*Example 2.3*

Consider again the rolling of a fair die, and define a random variable

$$X = \begin{cases} 1 & i = 6 \\ 0 & i \neq 6 \end{cases} \tag{2.89}$$

where $i$ is the number of dots on each face of the die. We shall find the cdf and pmf of this random variable, and then find its generating function. Finally, we shall find the generating function of a sum of $n$ such, independent, random variables, and so determine the pmf of the latter.

Let

$$P(X = 1) = p \tag{2.90}$$

and

$$P(X = 0) = q = 1 - p \tag{2.91}$$

To find the cdf, $F(x)$, for every $x$ from $-\infty$ to $\infty$, then:
If $x \geq 1$, then $X = 1 \leq x$ and $X = 0 \leq x$.

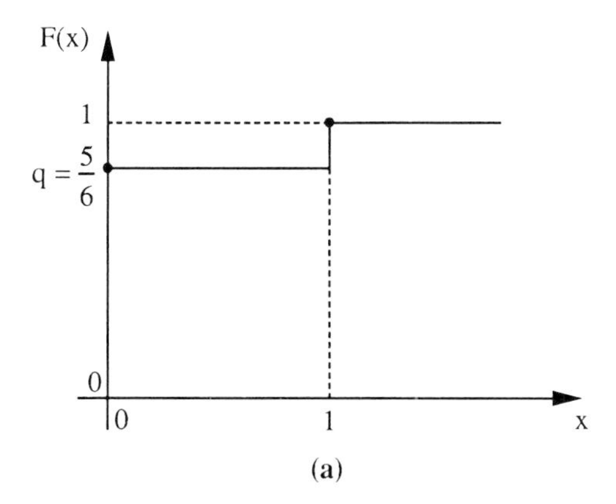


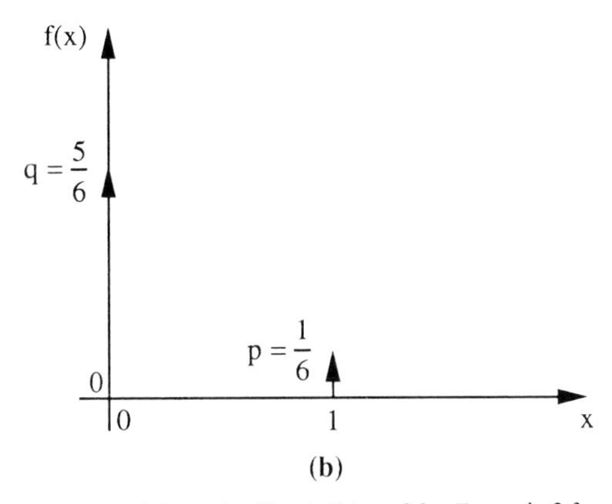


*Figure 2.4 (a) cdf and (b) pmf for Example 2.3*

Thus

$$F(x) = P(X \leq x) = p + q = 1; \qquad x \geq 1 \tag{2.92}$$

If $0 \leq x < 1$, then $X = 1 > x$ and $X = 0 > x$. Hence

$$F(x) = P(X \leq x) = q; \qquad 0 \leq x < 1 \tag{2.93}$$

If $x < 0$, then $X = 1 > x$ and $X = 0 > x$. Hence

$$F(x) = P(X \leq x) = P(\phi) = 0; \qquad x < 0 \tag{2.94}$$

Also, since $p$ and $q$ have values of 1/6 and 5/6, respectively, then the cdf and pmf have the forms shown in Fig. 2.3(a) and Fig. 2.3(b), respectively.

The generating function for this distribution is given by (2.76) as

$$p(z) = \sum_{i=0}^{\infty} p_i z^i$$

where

$$p_0 = P(X = 0) = q \tag{2.95}$$

and

$$p_1 = P(X = 1) = p \tag{2.96}$$

Thus using (2.95) and (2.96) in (2.76)

$$\begin{aligned} p(z) &= q + pz \\ &= 1 - p(1 - z) \end{aligned} \tag{2.97}$$

If we have $n$ such random variables that are independent, then the generating function of their sum is given by equation (2.88) as

$$r(z) = [1 - p(1 - z)]^n \tag{2.98}$$

To find the pmf of this new distribution, we could proceed by using (2.98), above, in conjunction with equation (2.78) to evaluate the individual probabilities $p_k$, $k = 0, 1, \ldots$, where $p_k = P(S = k)$, and $S$ is the new random variable that has been formed by summing the $n$ originals. This can be done more conveniently, however, by simply expanding the right hand side of (2.98) in powers of $z$. The desired probability, $p_k$, is then obtained by picking out the coefficient of $z^k$. Clearly, from the Binomial Theorem this coefficient is simply

$$p_k = \binom{n}{k} p^k q^{n-k} \tag{2.99}$$

where

$$\binom{n}{k} = \frac{n!}{(n - k)!k!}$$

The probability mass function we have derived in (2.99) is that of a random variable that is said to have (for rather obvious reasons) a *binomial distribution.* Recall that this new random variable was formed by summing $n$ independent random variables whose individual distributions depended on the success or failure of a single trial (in this case rolling a die, where a six was counted as a success, and any other number, a failure). Such trials are called *Bernoulli trials*, and the associated random variables are called *Bernoulli variables.* We see, therefore, that the binomial pmf gives the probability of $k$ successes in $n$ independent trials, $0 \leq k \leq n$, where each trial results in a success with probability $p$, and a failure with probability $q = 1 - p$. Thus, in the context of Example 2.3, equation (2.99) gives the probability, $p_k$ of turning up exactly $k$ sixes, in $n$ rolls of a fair die.

So far we have considered generating functions only for discrete random variables. Similar functions can, of course, be defined for continuous random variables, but using integrals instead of sums. In the remainder of the text, however, we shall not have any call for their use in the context of continuous random variables and so the topic will be skipped. The interested reader can find details of generating functions for continuous random variables in almost any textbook on probability theory; for example, [FELL 71], [PAPO 84], or [ROSS 88].

## 2.3 DISTRIBUTIONS

In this section, we consider some specific distributions of random variables, where we are here using the term 'distribution' in a loose sense, to mean both the cdf and the pdf (or pmf in the discrete case). The treatment is not intended to be comprehensive, and we consider only those distributions that are either directly encountered later in the text, or are necessary to its understanding.

### 2.3.1 The uniform random variable

Consider a continuous random variable, $X$, that is equally likely to take any value in a specified range. Such a random variable is said to be *uniformly distributed* on its range, where the range of values is usually an interval on the real line; thus $a < x < b$. If all values of this random variable are to be equally likely, then the pdf of $X$ must be a constant, $c$, on the interval $(a, b)$, and 0 elsewhere. Thus

$$f(x) = \begin{cases} c & \text{if } a < x < b \\ 0 & \text{elsewhere} \end{cases} \tag{2.100}$$

For the normalising condition of equation (2.50) to be satisfied:

$$\int_{-\infty}^{\infty} f(x)\, dx = 1$$

This results in

$$c = \frac{1}{b - a} \tag{2.101}$$

The cdf is obtained from (2.49) as

$$F(x) = \int_{-\infty}^{\infty} f(u)\, du$$

$$= \begin{cases} & 0 \text{ if } x \leq a \\ \left[\dfrac{x - a}{b - a}\right] & \text{if } a < x < b \\ & 1 \text{ if } b \leq x \end{cases} \tag{2.102}$$

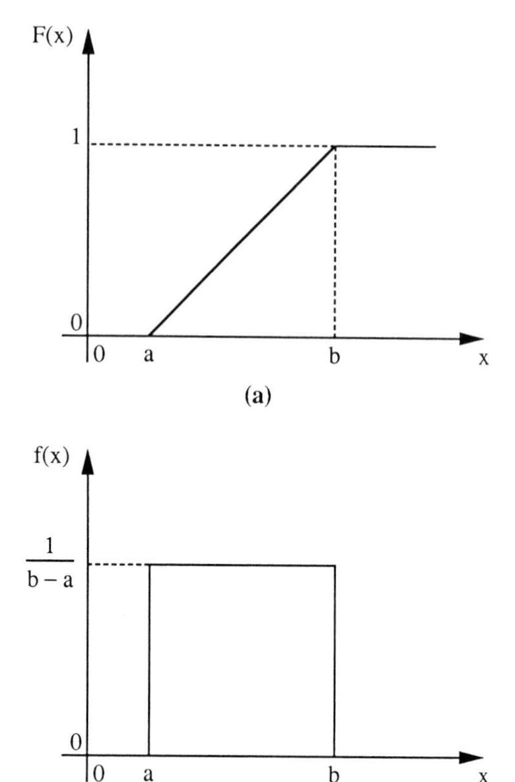


*Figure 2.5 (a) cdf and (b) pdf of a continuous random variable X*

The cdf and pdf for this random variable are shown in Fig. 2.5.

If $(d, e)$ is any sub-interval of $(a, b)$, then the probability that $X$ takes a value in this sub-interval depends only on the length of the sub-interval, and not on its position. Thus

$$P(d < X < e) = F(e) - F(d)$$
$$= \frac{e - d}{b - a} \tag{2.103}$$

The mean and variance of this distribution are, respectively

$$E[X] = \int_{-\infty}^{\infty} xf(x)\,dx$$

$$= c \int_a^b x \, dx$$

$$= \frac{a+b}{2} \tag{2.104}$$

and

$$\mathrm{Var}[X] = \int_{-\infty}^{\infty} (X - E[X])^2 f(x) \, dx$$

$$= E[X^2] - (E[X])^2$$

$$= c \int_a^b x^2 \, dx - \left(\frac{a+b}{2}\right)^2$$

$$= \frac{(b-a)^2}{12} \tag{2.105}$$

Of all the possible distribution functions on the interval $(a, b)$ the uniform one can be considered the most random in the sense that it offers the least help in predicting the value that the corresponding random variable will take.

It is worth noting in passing that the most widely used distribution function in the simulation modelling of communication or computer networks is the continuous random variable uniformly distributed on the interval (0, 1). There are two reasons for this: First, it can be used as a basis for generating any other distribution, whether continuous or discrete; second, the function itself can be easily implemented in the form of a linear congruential generator, using just a few lines of code from a high level programming language. Those interested in computer simulation modelling are referred to the excellent text by Matloff [MATL 88].

In the discrete case, the uniform random variable $X$ can take $n$ discrete values, say $a, a+1, a+2, \ldots, b = a+n-1$, where $a$ is usually an integer, each with probability

$$P(X = i) = p_i = \frac{1}{n}; \qquad i = a, a+1, a+2, \ldots, b \tag{2.106}$$

An example of this type of random variable is the number of dots that show on the face of a fair die when it is rolled, in which case $a = 1$ and $b = n = 6$. The cdf and pmf for this random variable are similar to those shown in Fig. 2.3, except that the $x$ axis should be divided by 10.

In general, the pmf is given by

$$f(x) = \begin{cases} \dfrac{1}{n} & \text{if } a \le x \le b \\ 0 & \text{elsewhere} \end{cases} \tag{2.107}$$

and the cdf is

$$F(x) = \frac{1}{n}\sum_{i=a}^{b} \delta(x - i) \tag{2.108}$$

Note that the staircase function of the cdf in Fig. 2.3(a) tends to the ramp function shown in Fig. 2.5(a) when $n \to \infty$. Likewise, the pmf of Fig. 2.3(b) tends to the pdf of Fig. 2.5(b) when $n \to \infty$. In other words, when the number of possible values of a discrete random variable tends to infinity, then the cdf and pmf of the discrete variable tend to their continuous counterparts.

We shall now evaluate the mean and variance of the discrete uniformly distributed random variable for the case $a = 1$ by the use of generating functions, in order to illustrate the use of the latter for this purpose.

Since the generating function is defined by

$$p(z) = \sum_{i=0}^{\infty} p_i z^i$$

Substituting in the above from equation (2.106), and setting $a = 1$, we get

$$\begin{aligned} p(z) &= \frac{1}{n}\sum_{i=1}^{n} z^i \\ &= \frac{1}{n}(z + z^2 + \cdots + z^n) \end{aligned} \tag{2.109}$$

Hence

$$p^{(1)}(z) = \frac{1}{n}(1 + 2z + \cdots + nz^{n-1}) \tag{2.110}$$

and so from equation (2.81), the mean value of $X$ is

$$\begin{aligned} E[X] &= p^{(1)}(1) \\ &= \frac{1}{n}\sum_{i=1}^{n} i \\ &= \frac{n+1}{2} \end{aligned} \tag{2.111}$$

To find the variance, we first note from equation (2.68) that

$$\mathrm{Var}[X] = E[X^2] - (E(X])^2$$

and from equation (2.83) that

$$\begin{aligned} E[X^2] &= p^{(2)}(1) + E[X] \\ &= p^{(2)}(1) + p^{(1)}(1) \end{aligned} \tag{2.112}$$

Then clearly

$$\mathrm{Var}[X] = p^{(2)}(1) + p^{(1)}(1)[1 - p^{(1)}(1)] \tag{2.113}$$

Thus differentiating equation (2.110) with respect to $z$ we have

$$\begin{aligned} p^{(2)}(z) &= \frac{1}{n}(2 + 3.2z + 4.3z^2 + \cdots + n(n-1)z^{n-2}) \\ &= \frac{1}{n}\sum_{i=1}^{n-1} i^2 z^{i-1} + \frac{1}{n}\sum_{i=1}^{n-1} i z^{i-1} \end{aligned} \tag{2.114}$$

This gives

$$\begin{aligned} p^{(2)}(1) &= \frac{1}{n}\sum_{i=1}^{n-1} i^2 + \frac{1}{n}\sum_{i=1}^{n-1} i \\ &= \frac{(2n-1)(n-1)}{6} + \frac{(n-1)}{2} \\ &= \frac{(n^2-1)}{3} \end{aligned} \tag{2.115}$$

Finally, using equations (2.111) and (2.115) in (2.113), we have

$$\mathrm{Var}[X] = \frac{n^2 - 1}{12} \tag{2.116}$$

It is not difficult to generalise the results of equations (2.111) and (2.116) for any specific range, $a, a+1, a+2, \ldots, b = a + n - 1$, in which case we get

$$E[X] = \frac{(a+b)}{2} \tag{2.117}$$

and

$$\mathrm{Var}[X] = \frac{(b - a + 2)(b - a)}{12} \tag{2.118}$$

Comparing these results with those of the equivalent continuous uniform random variable, given by equations (2.104) and (2.105), it is interesting to note that the mean values are the same, obviously due to the symmetry of both distributions, but discretising the (finite) sample space causes the variance to increase.

### 2.3.2 The exponential random variable

A continuous random variable, $X$, is said to be *exponentially distributed*, with *parameter* $\lambda$ ($\lambda > 0$), if it has a cdf of the form:

$$F(x) = 1 - e^{-\lambda x}; \qquad x \geq 0 \tag{2.119}$$

The corresponding pdf is simply the derivative of this with respect to $x$ (see (2.47)), and so

$$f(x) = \lambda e^{-\lambda x}; \qquad x \geq 0 \tag{2.120}$$

The mean value of this random variable is

$$\begin{aligned} E[X] &= \int_0^{\infty} x\lambda e^{-\lambda x}\, dx \\ &= \frac{1}{\lambda} \end{aligned} \tag{2.121}$$

Thus we see that the parameter $\lambda$ is the reciprocal of the mean of the corresponding random variable.

The most important feature of an exponentially distributed random variable is its *memoryless property*. More precisely, if a random variable $X$ is distributed exponentially, then

$$P(X \leq t + x | X > t) = P(X \leq x); \qquad t, x \geq 0 \tag{2.122}$$

In other words, if we interpret the random quantity as time, then knowing that an exponentially distributed activity has been in progress for a time $t$ does not affect the distribution of its remaining duration – it is as if the process is starting *now*.

Equation (2.122) is quite easy to prove. Using the conditional probability formula, (2.23), then

$$\begin{aligned} P(X \leq t + x | X > t) &= \frac{P(t < X \leq t + x)}{P(X > t)} \\ &= \frac{F(t + x) - F(t)}{1 - F(t)} \\ &= \frac{e^{-\lambda t} - e^{-\lambda(t+x)}}{e^{-\lambda t}} \\ &= 1 - e^{-\lambda x} \\ &= P(X \leq x) \end{aligned}$$

It can be shown that the exponential distribution is the only continuous distribution that has the memoryless property. This property is also called the *Markov* property, and, as will be seen, it has far reaching consequences in the modelling of communication or computer networks.

The cdf and pdf of an exponentially distributed random variable, $X$, with parameter $\lambda$, are shown in Fig. 2.6. Note that for the pdf, the value

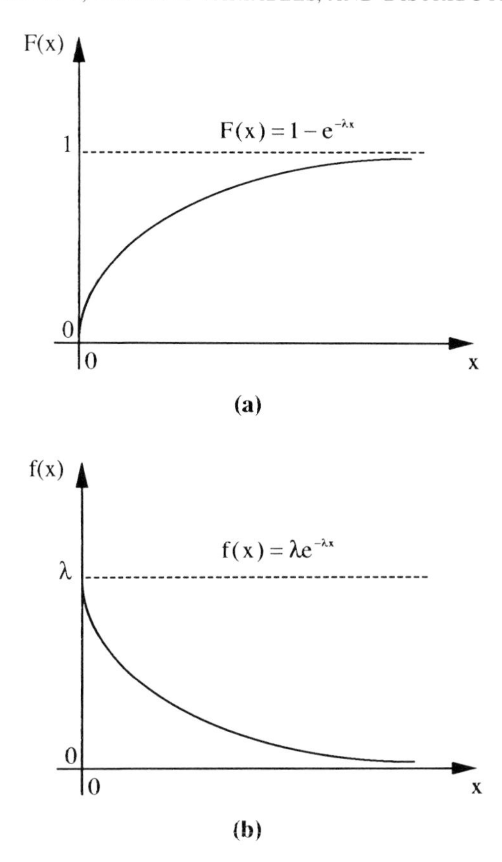


*Figure 2.6 (a) cdf and (b) pdf of an exponential random variable, X, with parameter λ*

of $\lambda$ defines the intercept on the $f(x)$ axis. Since the area under the pdf curve must be unity, then $\lambda$ also defines the rate of decay of the pdf curve. For this reason, $\lambda$ is also referred to as the *rate* of the distribution.

### 2.3.3 The geometric distribution

The discrete equivalent of the exponential distribution is the *geometric distribution*. A discrete random variable $X$ is said to be geometrically distributed if it has a pmf defined by

$$\begin{aligned} f(x) &= P(X = x) \\ &= q^{x-1}p \end{aligned} \tag{2.123}$$

where

$$q = 1 - p$$

and

$$x = 1, 2, \ldots$$

The random variable, $X$, can be interpreted as the index, $x$, of the *first* success in an experiment consisting of a (possibly infinite) number of independent trials, where each trial results in a success with probability $p$, and a failure with probability $q = 1 - p$ (Bernoulli trials). The cdf of $X$ is given by

$$\begin{aligned} F(x) &= P(X \leq x) \\ &= \sum_{i=1}^{x} p_i \\ &= p \sum_{i=1}^{x} q^{i-1} \\ &= 1 - q^x; \qquad x = 1, 2, \ldots \end{aligned} \tag{2.124}$$

The cdf is thus obtained from the sum of a geometric series, a characteristic from which the distribution takes its name.

In the context of rolling a fair die, we might construct an experiment such that a success occurs when an even number is rolled, and the die is to be rolled until a success occurs. The random variable, $X$, representing the index of this first success is thus geometrically distributed with $p = q = 1/2$. The cdf and pmf of this random variable are shown in Fig. 2.7.

The generating function of the random variable, $X$, can be evaluated as

$$\begin{aligned} p(z) &= \sum_{i=1}^{\infty} pq^{i-1}z^i \\ &= \frac{p}{q} \sum_{i=1}^{\infty} (qz)^i \\ &= \frac{p}{q}\left[\frac{1}{1 - qz} - 1\right] \\ &= \frac{pz}{1 - qz} \end{aligned} \tag{2.125}$$

From this the first and second order factorial moments of $X$, $p^{(1)}(1)$ and $p^{(2)}(1)$ respectively, can be obtained, and the mean and variance of $X$ can be evaluated using the method given in Section 2.3.2. This results in

$$E[X] = \frac{1}{p} \tag{2.126}$$

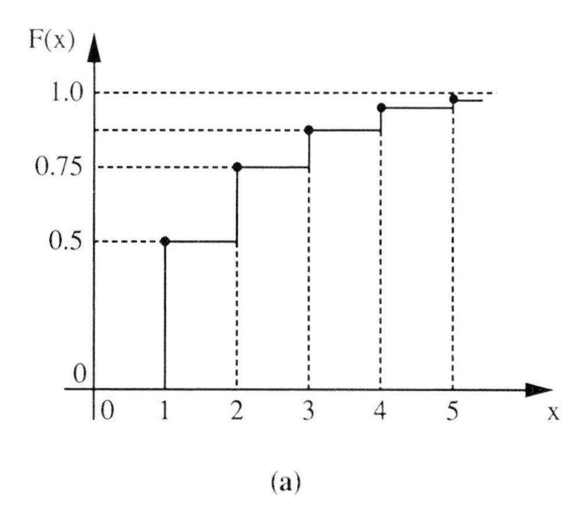


(a)

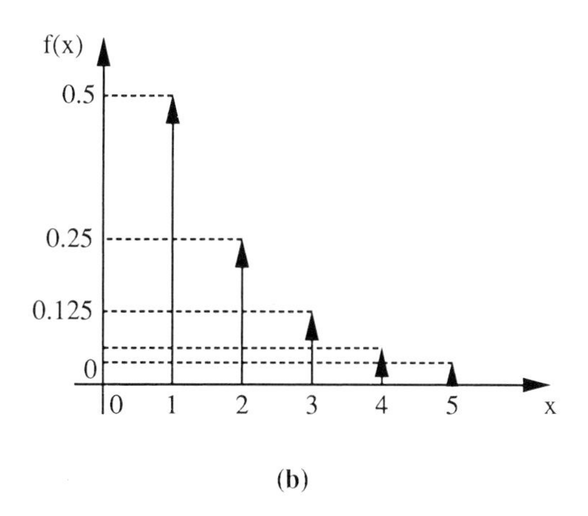


(b)

*Figure 2.7 (a) cdf and (b) pmf for a geometrically distributed random variable, X, with p = 1/2*

and

$$\mathrm{Var}[X] = \frac{(1-p)}{p^2} \tag{2.127}$$

Just as the exponential distribution is the only continuous distribution that has the memoryless property, its discrete equivalent, the geometric distribution, is the only discrete distribution that has this same property. Because of this, as will later be seen, the geometric distribution is of profound importance in discrete-time stochastic modelling, in particular where Markov chains are employed.

### 2.3.4 The binomial distribution

We say that a random variable, $X$, has a *binomial* distribution of order $n$ if it takes the values $0, 1, \ldots, n$ with probability

$$P(X = k) = p_k = \binom{n}{k} p^k q^{n-k} \tag{2.128}$$

and

$$p + q = 1$$

We have already met this distribution in Section 2.2.2 where it was derived as a sum of $n$ independent Bernoulli random variables.

The pmf is a sum of impulses

$$f(x) = \sum_{k=0}^{n} \binom{n}{k} p^k q^{n-k} \delta(n - k) \tag{2.129}$$

and the corresponding cdf is a staircase function in the interval $[0, n]$ given by

$$F(x) = \sum_{k=0}^{m} \binom{n}{k} p^k q^{n-k} \qquad m \leq x < m + 1 \tag{2.130}$$

The generating function of this random variable has already been derived (Equation 2.98) as

$$r(z) = [1 - p(1 - z)]^n$$

From this the mean and variance are easily obtained as

$$E[X] = np \tag{2.131}$$

and

$$\text{Var}[X] = np(1 - p) \tag{2.132}$$

The binomial random variable is closely related to the geometric random variable, but in this case the index, $k$, represents the number of successes in $n$ independent trials, where each trial has a probability $p$ of success, and probability $q = 1 - p$ of failure.

The binomial distribution is of fundamental importance in any discrete-time modelling situation, and from it can be obtained other distributions as limiting cases. Two of these, the *Poisson* distribution and the *Normal* (or *Gaussian*) distribution will next be discussed.

### 2.3.5 The Poisson distribution

A random variable, $X$, taking on one of the values $0, 1, 2, \ldots,$ is said to be a *Poisson* random variable with parameter $\lambda$ if for some $\lambda > 0$

$$P(X = k) = p_k = \frac{\lambda^k}{k!}e^{-\lambda} \qquad k = 0, 1, 2, \ldots \tag{2.133}$$

Equation (2.133) defines a pmf, since

$$\begin{aligned}\sum_{k=0}^{\infty} p_k &= e^{-\lambda} \sum_{k=0}^{\infty} \frac{\lambda^k}{k!} \\ &= e^{-\lambda}e^{\lambda} \\ &= 1\end{aligned}$$

The pmf may thus be written as

$$f(x) = e^{-\lambda} \sum_{k=0}^{\infty} \frac{\lambda^k}{k!}\delta(x - k) \tag{2.134}$$

and the cdf as

$$F(x) = e^{-\lambda} \sum_{k=0}^{m} \frac{\lambda^k}{k!} \qquad m \leq x < m + 1 \tag{2.135}$$

This is known as the Poisson distribution, and has a tremendous range of applications in many diverse areas. That it is a limiting case of the binomial distribution when $n$ is large and $np$ is of moderate size, such that $np = \lambda$, can easily be shown as follows.

Suppose that $X$ is a binomial random variable such that $np = \lambda$. Then

$$\begin{aligned}P(X = k) &= \frac{n!}{(n-k)!k!}p^k(1-p)^{n-k} \\ &= \frac{n!}{(n-k)!k!}\left(\frac{\lambda}{n}\right)^k\left(1 - \frac{\lambda}{n}\right)^{n-k} \\ &= \frac{n(n-1)\ldots(n-k+1)}{n^k}\frac{\lambda^k}{k!}\frac{\left(1 - \frac{\lambda}{n}\right)^n}{\left(1 - \frac{\lambda}{n}\right)^k}\end{aligned}$$

Now, for $n$ large and $\lambda$ moderate

$$\left(1 - \frac{\lambda}{n}\right)^n \simeq e^{-\lambda}$$

$$\frac{n(n-1)\ldots(n-k+1)}{n^k} \simeq 1$$

and

$$\left(1 - \frac{\lambda}{n}\right)^k \simeq 1$$

Then we have

$$P(X = k) \simeq e^{-\lambda}\frac{\lambda^k}{k!}$$

The generating function of this random variable is given by

$$p(z) = e^{\lambda(z-1)} \tag{2.136}$$

from which the mean and variance are found to be equal and given by

$$E[X] = \mathrm{Var}[X] = \lambda \tag{2.137}$$

The Poisson random variable will be discussed in more detail later in the context of a stochastic process. Perhaps its most frequent use is in modelling the arrival process of information units to a system or network, where the parameter $\lambda$ can be interpreted as the mean number of arrivals per unit time, in which case the probability $P(X = k)$ gives the probability of $k$ arrivals per unit time.

### 2.3.6 The normal distribution

If in the binomial distribution both $n$ and the expected value, $np$, are large, then in this case the corresponding random variable, $X$, can be approximated by one that has a pdf defined by

$$f(x) = \frac{1}{\sigma\sqrt{2\pi}} e^{\frac{-(x-\eta)^2}{2\sigma^2}} \qquad -\infty < x < \infty \tag{2.138}$$

This is the familiar *normal* (or *Gaussian*) density function, with mean value $\eta$ and variance $\sigma^2$; that is, the pmf of a binomial random variable becomes more and more normal as $n$ becomes larger and larger. Since, as we know, a binomial random variable is a sum of $n$ independent Bernoulli random variables, this implies that the distribution of the sum of a large number of independent Bernoulli random variables is approximately normal. Indeed, a very much stronger result than this, which is quite remarkable, is the *Central Limit Theorem.*

The Central Limit Theorem states that if $n$ random variables, $X_1, X_2, \ldots, X_n$, are independent, then under general conditions, the density $f(x)$ of their sum

$$X = X_1 + X_2 + \cdots + X_n \tag{2.139}$$

properly normalized, tends to a normal curve as $n \to \infty$. In other words, if $n$ is sufficiently large, then no matter what the distribution of the individual random variables $X_i$, $i = 1, 2, \ldots, n$, the distribution of their sum can be approximated by

$$f(x) \simeq \frac{1}{\sigma\sqrt{2\pi}} e^{\frac{-(x-\eta)^2}{2\sigma^2}} \tag{2.140}$$

If the random variables are discrete and $X$ takes values $ka$ with probability $P(X = ka) = p_k$, then the probabilities $p_k$ equal the samples of a normal curve

$$P(X = ka) \simeq \frac{1}{\sigma\sqrt{2\pi}} e^{\frac{-(ka-\eta)^2}{2\sigma^2}} \tag{2.141}$$

Since $f(x)$ is the convolution of the densities $f_i(x)$ of the random variables $X_i$, then

$$f(x) = f_1(x) * f_2(x) * \cdots * f_n(x) \tag{2.142}$$

Clearly, the Central Limit Theorem can also be viewed as a property of the convolutions of positive functions.

## 2.4 CONDITIONAL DISTRIBUTIONS

The *conditional distribution*, $F(x|M)$, of a random variable, $X$, given $M$, is defined as the conditional probability of the event $(X \leq x)$

$$F(x|M) = P(X \leq x|M) = \frac{P(X \leq x, M)}{P(M)} \tag{2.143}$$

where $(X \leq x, M)$ is the intersection of the events $X \leq x$ and $M$. Thus $F(x|M)$ has the same properties as $F(x)$, as given by equation (2.42), provided that probabilities are replaced by conditional probabilities. In particular

$$F(\infty|M) = 1 \qquad F(-\infty|M) = 0 \tag{2.144}$$

$$P(x_1 < X \leq x_2|M) = F(x_2|M) - F(x_1|M) = \frac{P(x_1 < X \leq x_2, M)}{P(M)} \tag{2.145}$$

The *conditional density* $f(x|M)$ is the derivative of $F(x|M)$

$$f(x|M) = \frac{dF(x|M)}{dx} \tag{2.146}$$

This function is non-negative, and its area equals 1.

As an example, if $M = (b < x \leq a)$ then equation (2.143) gives

$$F(x|b < X \leq a) = \frac{P(X \leq x, b < x \leq a)}{P(b < x \leq a)} \tag{2.147}$$

If $x \geq a$, then $(X \leq x, b < x \leq a) = (b < X \leq a)$. Hence

$$F(x|b < X \leq a) = \frac{F(a) - F(b)}{F(a) - F(b)} = 1 \tag{2.148}$$

If $b \leq x < a$, then $(X \leq x, b < x \leq a) = (b < X \leq x)$. Hence

$$F(x|b < X \leq a) = \frac{F(x) - F(b)}{F(a) - F(b)} \tag{2.149}$$

If $x < b$, then $(X \leq x, b < x \leq a) = (\phi)$. Hence

$$F(x|b < X \leq a) = 0 \tag{2.150}$$

The density is

$$f(x|b < X \leq a) = \frac{f(x)}{F(a) - F(b)} \qquad \text{for } b \leq x < a \tag{2.151}$$

and zero otherwise. This is shown in Fig. 2.8.

Clearly then, a pdf, $f(x)$, of a random variable $X$, when conditioned on an interval $b < x \leq a$ is effectively scaled by $1/(F(a) - F(b))$, where $F(x)$ is the cdf of $X$. The result, of course, also carries through when $X$ is a discrete random variable. As an illustration of this, consider the following example.

*Example 2.4*

An urn contains a large number of coloured balls, a fraction $p$ of which are white. A second urn contains $N$ white balls, and a third urn is empty but has room for exactly $M$ balls, $M \leq N$. A total of $N$ balls are withdrawn

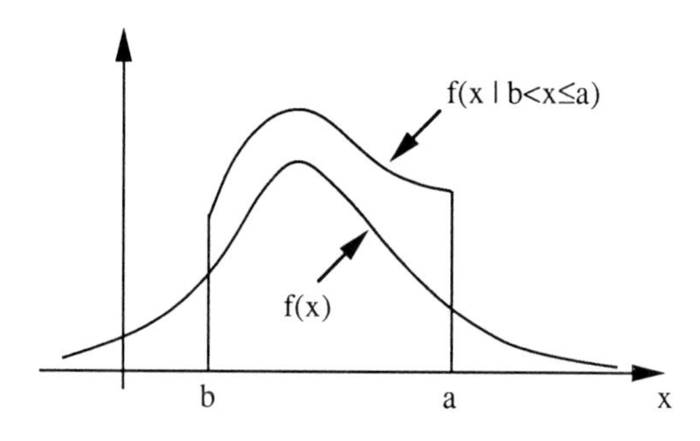


*Figure 2.8 A conditional density function*

one at a time, and immediately replaced, from the first urn. Each time a white ball is selected, a white ball is transferred from the second urn to the third urn. If the third urn is found to be full then the ball is simply replaced in the second urn. If $X$ is a random variable representing the number of balls in the third urn at the end of the experiment, find its mean value.

Clearly, $X$ is a binomial random variable conditioned on $0 \leq X \leq M$. According to equation (2.151) we have

$$f(x|0 \leq X \leq M) = \frac{f(x)}{F(M) - 0} \tag{2.152}$$

Thus

$$P(X = x) = \frac{\binom{N}{x} p^x (1-p)^{N-x}}{\sum_{i=0}^{M} \binom{N}{i} p^i (1-p)^{N-i}} \quad \text{for } x = 0, 1, \ldots, M \tag{2.153}$$

We now require to find

$$E[X] = p^{(1)}(1) \tag{2.154}$$

where

$$p(z) = \sum_{i=0}^{M} P(X = i) z^i$$

Direct calculation of $p^{(1)}(z)$ is not straightforward, and instead we define a second function

$$g(z) = \sum_{i=0}^{M} P(X = i) z^{N-i} \tag{2.155}$$

The trick is now to focus on calculating $g^{(1)}(z)$ rather than $p^{(1)}(z)$. To this end we have

$$p^{(1)}(z) = \sum_{i=0}^{M} P(X = i) i z^{i-1}$$

$$= \frac{1}{F(M)} \sum_{i=0}^{M} \binom{N}{i} p^i (1-p)^{N-i} i z^{i-1} \tag{2.156}$$

where

$$F(M) = \sum_{i=0}^{M} \binom{N}{i} p^i (1-p)^{N-i}$$

Also

$$g^{(1)}(z) = \sum_{i=0}^{M} P(X = i)(N - i)z^{N-1-i}$$

$$= \frac{1}{F(M)} \sum_{i=0}^{M} \binom{N}{i} p^i (1 - p)^{N-i} (N - i) z^{N-1-i} \tag{2.157}$$

Thus, replacing $z$ by 1 in equations (2.156) and (2.157), this leads to

$$g^{(1)}(1) = N - p^{(1)}(1) \tag{2.158}$$

Now, equation (2.157) can be written as

$$g^{(1)}(z) = \frac{1}{F(M)} \sum_{i=0}^{M} \frac{N!}{(N - 1 - i)!\,i!} p^i (1 - p)^{N-i} z^{N-1-i} \tag{2.159}$$

Writing $j = i + 1$, then

$$g^{(1)}(z) = \frac{1}{F(M)} \frac{(1 - p)}{p} \sum_{j=1}^{M} j \binom{N}{j} p^j (1 - p)^{N-j} z^{N-j}$$
$$+ \frac{1}{F(M)} \frac{(1 - p)}{p} \binom{N}{M + 1} (M + 1) p^{M+1} (1 - p)^{N-1-M} z^{N-1-M} \tag{2.160}$$

Using equation (2.156) we thus have

$$g^{(1)}(1) = \frac{1 - p}{p} p^{(1)}(1) + \frac{M + 1}{F(M)} \binom{N}{M + 1} p^M (1 - p)^{N-M} \tag{2.161}$$

Finally, using equation (2.158), then

$$p^{(1)}(1) = Np - \frac{(1 - p)(M + 1)}{F(M)} \binom{N}{M + 1} p^{M+1} (1 - p)^{N-(M+1)} \tag{2.162}$$

Equation (2.162) thus gives the mean value of a binomial random variable, conditioned on the interval $[0, M]$. Note that the effect of conditioning is to reduce the mean value of the standard (unconditioned) binomial random variable by an amount equal to the second term on the right hand side of equation (2.162). If $p$ is small, then except for the trivial case when $M = 0$ this second term will also be small compared to $Np$, and the mean value can be approximated by that of the unconditioned binomial random variable. That is, we are cutting off a very small tail of the unconditioned binomial mass function.

A conditional binomial distribution of the type examined above will be later found to have extensive use in the modelling of slotted ring type networks [PIER 72] in the sense that if such a system has $N$ users and $M$ slots, $N \geq M$, then not more than $M$ slots can be filled by messages from the $N$ users at any given time.

# 3. STOCHASTIC PROCESSES AND MARKOV CHAINS

*Chapter Objectives: To cover the basic principles of discrete-time Markov chains, and show how these can be used to model the performance of a communication or computer network.*

A stochastic process is a random process that changes with time. More formally, a stochastic process is defined as a set of random variables $\{x_t, t \in T\}$ defined on the same sample space $T$. The set $T$ is usually interpreted as a set of time instants. When $t$ is countable we have a *discrete-time stochastic process*, and when the elements of $T$ are a continuous set of real numbers, we then have a *continuous-time stochastic process*. $x_t$ is called the *state* of the process at time $t$. A *sample path* is a single instance of a stochastic process, and any arguments that apply to a sample path also apply to the process as a whole. A stochastic process is said to be *stationary* if the distribution of $x_{t+s} - x_t$ is independent of $t$. In the following we shall look at two important stochastic processes, a *Poisson* process and a special case of a *Markov process* called a *Markov chain.*

## 3.1 POISSON PROCESS

The Poisson process is one of the simplest of all stochastic processes and is frequently used to define the *arrival process* of information units (messages) at a user in a communication or computer network. As its name implies, it is based on the Poisson distribution, and, interestingly it is a purely random arrival process. One way of looking at it goes like this. Suppose the time axis is divided into a large number of segments of width $\Delta t$. Let the probability of a single message arriving in a segment be proportional to the width of the segment, $\Delta t$, with a proportionality constant, $\lambda$, which represents the mean arrival rate.

$$P\ (\textit{exactly } 1 \textit{ arrival in } (t, t + \Delta t)) = \lambda \Delta t \tag{3.1}$$

$$P\ (\textit{no arrivals in } (t, t + \Delta t)) = 1 - \lambda \Delta t \tag{3.2}$$

$$P\ (\textit{more than one arrival in } (t, t + \Delta t)) = 0 \tag{3.3}$$

The last equation implies that we ignore higher order terms in $\Delta t$.

We can thus consider each segment to contain a Bernoulli trial, with $\lambda \Delta t$ the probability of success (a message arrival) and $1 - \lambda \Delta t$ the probability of failure (no arrival).

As $\Delta t \to 0$ we form the continuous time Poisson process. Clearly, the arrivals are independent of one another, and no one interval is any more or less likely to have an arrival than any other interval.

We will derive the underlying differential equation of the Poisson process by using difference equation arguments and letting $\Delta t \to 0$. Begin by defining

$$P_n(t) \equiv P\ (\textit{total arrivals} = n \textit{ at time } t) \tag{3.4}$$

and let $p_{ij}(\Delta t)$ be the probability of going from $i$ arrivals to $j$ arrivals in a time interval $\Delta t$ seconds. The following approach now follows [KLEI 75].

The number of arrivals is the state of the system and contains all the information to completely describe the system. We can write

$$P_n(t + \Delta t) = P_n(t)p_{n,n}(\Delta t) + P_{n-1}(t)p_{n-1,n}(\Delta t) \tag{3.5}$$

We have again neglected higher order terms in $\Delta t$. What this equation says is that one can arrive at a situation with $n$ messages arrived by time $t + \Delta t$ by having either $n$ or $n - 1$ messages by time $t$.

To complete the difference equation description we need a special equation for state 0 which expresses a boundary condition. This is

$$P_0(t + \Delta t) = P_0(t)p_{0,0}(\Delta t) \tag{3.6}$$

Now using equations (3.1), (3.2) and (3.3)

$$P_n(t + \Delta t) = P_n(t)(1 - \lambda \Delta t) + P_{n-1}(t)(\lambda \Delta t) \tag{3.7}$$

$$P_0(t + \Delta t) = P_0(t)(1 - \lambda \Delta t) \tag{3.8}$$

Multiplying out these expressions and rearranging, one arrives at

$$\frac{P_n(t + \Delta t) - P_n(t)}{\Delta t} = -\lambda P_n(t) + \lambda P_{n-1}(t) \tag{3.9}$$

$$\frac{P_0(t + \Delta t) - P_0(t)}{\Delta t} = -\lambda P_0(t) \tag{3.10}$$

If we let $\Delta t \to 0$, the above difference equations become differential equations:

$$\frac{dP_n(t)}{dt} = -\lambda P_n(t) + \lambda P_{n-1}(t) \tag{3.11}$$

$$\frac{dP_0(t)}{dt} = -\lambda P_0(t) \tag{3.12}$$

where $n \geq 1$. Equation (3.12) has a solution

$$P_0(t) = e^{-\lambda t} \tag{3.13}$$

Using this in equation (3.11), with $n = 1$

$$\frac{dP_1(t)}{dt} = -\lambda P_1(t) + \lambda e^{-\lambda t} \tag{3.14}$$

This equation has the solution

$$P_1(t) = \lambda t e^{-\lambda t} \tag{3.15}$$

Next, using $n = 2$ in equation (3.11)

$$\frac{dP_2(t)}{dt} = -\lambda P_2(t) + \lambda^2 t e^{-\lambda t} \tag{3.16}$$

This equation has the solution

$$P_2(t) = \frac{\lambda^2 t^2}{2} e^{-\lambda t} \tag{3.17}$$

Continuing this process, then by induction we find

$$P_n(t) = \frac{(\lambda t)^n}{n!} e^{-\lambda t} \tag{3.18}$$

We thus have arrived at the Poisson distribution, previously derived as a limiting case of the binomial distribution.

Note the subtle difference between the *rate* of the process, which is $\lambda$ (arrivals per unit time), and the *parameter* of the process, which is $\lambda t$. As shown previously, $\lambda t$ is also the mean and the variance.

## 3.2 PROPERTIES OF THE POISSON PROCESS

### 3.2.1 Union property

Consider two arrival processes, $A_1$ and $A_2$, which are independent and Poisson with rates $\lambda_1$ and $\lambda_2$ respectively. What can be said about the merger, or superposition $B$, of the two processes?

Let $I$ be an interval, or a union of a finite number of disjoint intervals, of length $t$, and let $K_1$, $K_2$ and $K$ be the numbers of arrivals in $I$, associated with $A_1$, $A_2$ and $B$, respectively. We know that $K_1$ and $K_2$ are independent and their distributions are Poisson, with parameters $\lambda_1 t$ and $\lambda_2 t$ respectively. Using equation (2.136) the generating functions of $K_1$ and $K_2$ are, respectively

$$p_{K_1}(z) = e^{\lambda_1 t(z-1)} \tag{3.19}$$

and

$$p_{K_2}(z) = e^{\lambda_2 t(z-1)} \tag{3.20}$$

Thus

$$p_K(z) = e^{(\lambda_1+\lambda_2)t(z-1)} \tag{3.21}$$

This is the generating function of a Poisson random variable with rate $\lambda_1 + \lambda_2$. Therefore $K$ has the Poisson distribution with parameter $(\lambda_1 + \lambda_2)t$. This, in turn, implies that the superposition process $B$ is Poisson with rate $\lambda_1 + \lambda_2$.

The same argument applied to the merger of more than two processes yields the *superposition property*, stated as follows:

If $A_1, A_2, \ldots, A_n$ are independent Poisson processes with rates $\lambda_1, \lambda_2, \ldots, \lambda_n$ respectively, then their superposition is also a Poisson process with rate $\lambda_1 + \lambda_2 + \cdots + \lambda_n$.

### 3.2.2 Decomposition property

Consider next the operation of splitting, or decomposing, a Poisson arrival process $A$, with rate $\lambda$, into two arrival processes $B_1$ and $B_2$.

The decomposition is done by a sequence of Bernoulli trials, where every arrival to the process $A$ is assigned to process $B_1$ with probability $p_1$ and is assigned to process $B_2$ with probability $p_2$, where $p_1 + p_2 = 1$, regardless of all previous assignments. Let $K$, $K_1$ and $K_2$ be the number of arrivals associated with $A$, $B_1$ and $B_2$, respectively, during an interval or a set of disjoint intervals, of length $t$. We know that $K$, which is equal to $K_1 + K_2$ has the Poisson distribution with parameter $\lambda t$; the problem is to find the joint distribution of $K_1$ and $K_2$. Now if the value of $K$ is given, then $K_1$ and $K_2$ are the number of successes and failures in the corresponding set of Bernoulli trials. Then:

$$\begin{aligned} P(K_1 = k_1, K_2 = k_2) & \\ &= P(K_1 = k_1, K_2 = k_2 | K = k_1 + k_2)P(K = k_1 + k_2) \\ &= \frac{(k_1+k_2)!}{k_1!k_2!} p_1^{k_1} p_2^{k_2} \cdot \frac{(\lambda t)^{k_1+k_2}}{(k_1+k_2)!} e^{-\lambda t} \\ &= \frac{(p_1 \lambda t)^{k_1}}{k_1!} e^{-p_1 \lambda t} \frac{(p_2 \lambda t)^{k_2}}{k_2!} e^{-p_2 \lambda t} \end{aligned} \tag{3.22}$$

Equation (3.22) implies that $K_1$ and $K_2$ are both Poisson distributions with parameters $p_1 \lambda t$ and $p_2 \lambda t$ respectively. Furthermore, these random variables are independent of each other. A generalisation of this argument yields the *decomposition property*, stated as follows:

If a Poisson process, $A$, with rate $\lambda$ is decomposed into processes $B_1, B_2, \ldots, B_n$, by assigning each arrival in $A$ to $B_i$ with probability $p_i$,

$i = 1, 2, \ldots, n$, and $p_1 + p_2 + \cdots + p_n = 1$, independent of all previous assignments, then $B_1, B_2, \ldots, B_n$ are Poisson processes with rates $p_1\lambda, p_2\lambda, \ldots, p_n\lambda$, respectively, and are independent of each other.

### 3.2.3 Interarrival times

The times between successive events in an arrival process are called *interarrival times.* To determine the distribution of the interarrival times in a Poisson process we can proceed as follows:

Let $X$ be a random variable representing the time between successive arrivals in a Poisson process. Then

$$\begin{aligned} P(X \leq t) &= 1 - P(X > t) \\ &= 1 - P_0(t) \\ &= 1 - e^{-\lambda t} \end{aligned} \tag{3.23}$$

Thus

$$F_X(t) = 1 - e^{-\lambda t} \tag{3.24}$$

Differentiating the above

$$f_X(t) = \lambda e^{-\lambda t} \tag{3.25}$$

Clearly, then, the interarrival times in a Poisson process of rate $\lambda$, by comparison with equations (2.119) and (2.120), are exponentially distributed with rate $\lambda$.

Because of the memoryless property of the exponential distribution, this in turn means that the interarrival times in a Poisson process are also memoryless; that is, if one takes a random point on the time axis, then the time to the next arrival is independent of the time since the last arrival. As previously mentioned in Section 2.3.2, the memoryless property is also called the Markov property, and has a profound significance in modelling the behaviour of communications and computer systems in that any stochastic process that has the Markov property is not dependent on the past in predicting its future behaviour. Such stochastic processes are called Markov processes, and in the following sections we shall examine a specific type of Markov process called a Markov chain.

## 3.3 MARKOV CHAINS

A *Markov process* is a type of stochastic process that is characterised by the so called *Markov property*. This property is the memoryless property that we found to be a characteristic of the exponential distribution, and hence also of the interarrival times of a Poisson process. The Poisson process is therefore a special case of a Markov process. In general, a

Markov process is a stochastic process such that if at a given time its state is known, then its subsequent behaviour is independent of its past history. There are a number of different types of Markov processes, and a good exposition of their interrelationships can be found in [KLEI 75]. In the following, we shall be concerned with a type of Markov process called a *Markov chain*.

The Markov chains that are of interest to us are *discrete-state, discrete-time Markov chains*. That is, the states can be numbered by the set, or by a subset, of the non-negative integers $\{0, 1, 2, \ldots\}$; similarly, time is discretised, with the time index, $n$, taking successive values as $\{n = 0, 1, 2, \ldots\}$. In all subsequent references to Markov chains it will be implicit that they are of the discrete-state, discrete-time form, unless otherwise stated.

### 3.3.1 Definitions and terminology

Let

$$\mathbf{x} = \{x_n = x; n = 0, 1, 2, \ldots\}, \qquad x \in \mathbf{X} \tag{3.26}$$

bet a Markov chain, where $x_n$ is the *state* at time $n$, and $\mathbf{X}$ is the *state space* defined by

$$\mathbf{X} = \{0, 1, 2, \ldots\} \tag{3.27}$$

The Markov property can be expressed as

$$P(x_{n+1} = j | x_0, x_1, \ldots, x_n) = P(x_{n+1} = j | x_n) \qquad j \in \mathbf{X}, n = 0, 1, 2, \ldots \tag{3.28}$$

That is, given the state at time $n$, then the state at time $n + 1$ is independent of the state at times $0, 1, \ldots, n - 1$.

Thus the evolution of a Markov chain is completely described by its *one-step transition probabilities*, $p_{ij}(n)$, that the chain will move to state $j$ at time $n + 1$, given that it is in state $i$ at time $n$:

$$p_{ij}(n) = P(x_{n+1} = j | x_n = i) \qquad i, j \in \mathbf{X}, n = 0, 1, 2, \ldots \tag{3.29}$$

We shall subsequently assume that these one-step transition probabilities do not depend on the time index, $n$:

$$p_{i,j}(n) = p_{ij}; \qquad i, j \in \mathbf{X}, n = 0, 1, 2, \ldots \tag{3.30}$$

Such Markov chains are *time-homogeneous*. This assumption simplifies the treatment without reducing the generality of the theory, and from now on it will be implicit that all Markov chains discussed are time-homogeneous.

It follows that a Markov chain is characterised by its *transition probability matrix*, $P$, defined by

$$P \equiv [p_{ij}, i, j \in \mathbf{X}]$$

$$= \begin{bmatrix} p_{00} & p_{01} & p_{02} & \cdot & \cdot \\ p_{10} & p_{11} & p_{12} & \cdot & \cdot \\ p_{20} & p_{21} & p_{22} & \cdot & \cdot \\ \cdot & \cdot & \cdot & \cdot & \cdot \\ \cdot & \cdot & \cdot & \cdot & \cdot \end{bmatrix} \quad (3.31)$$

This matrix is a *stochastic matrix*, which means that the elements in each row sum to 1. That is,

$$\sum_{j \in \mathbf{X}} p_{ij} = 1; \qquad i \in \mathbf{X} \quad (3.32)$$

An alternative way of representing the evolution of a Markov chain is a *state transition diagram*. This is a directed graph with vertices representing states, and arcs labelled with the corresponding one-step transition probabilities. To see this consider the following example.

*Example 3.1: Two interconnected users.*

Two users are connected via a single communication channel to a computer. Each has a continous stream of messages to send. Time is divided into slots, where each slot is sufficiently large to carry one fixed length message. The medium access control protocol is such that each user transmits a message with probability $p$ per slot, and defers to the next slot with probability $1 - p$. We see, therefore, that the channel can be in one of three states:

1. Idle – no user transmits in a slot.
2. Transmitting – exactly one of the two users transmits in a slot.
3. Collision – both users attempt transmission in the same slot.

If a message takes one slot to transmit, and the channel has zero propagation delay and is error-free, construct the state transition diagram of the Markov chain that describes the behaviour of the channel.

*Solution:* Define the (finite) state space as

$$\mathbf{X} = \{0, 1, 2\}$$

where state $0 =$ idle
$1 =$ transmission
$2 =$ collision

The one-step transition probability matrix is

$$P = \begin{bmatrix} p_{00} & p_{01} & p_{02} \\ p_{10} & p_{11} & p_{12} \\ p_{20} & p_{21} & p_{22} \end{bmatrix}$$

To determine these probabilities, we consider each state and its associated transition probabilities in turn. We shall assume that both users behave independently.

A little thought will soon show that each slot can be modelled as a binomial random variable with parameters $(2, p)$. That is, each slot can be considered as two independent trials, each having a probability $p$ of success (a transmission attempt).

We thus have for state 0:

$p_{00} = (1 - p)^2$; no transmission attempt, so the channel remains idle.

$p_{01} = 2p(1 - p)$; exactly one transmission attempt, so the channel moves to the transmit state.

$p_{02} = p^2$; two transmission attempts, so the channel moves to the collision state.

States 1 and 2 can be similarly considered to give

$$p_{10} = (1 - p)^2$$
$$p_{11} = 2p(1 - p)$$
$$p_{12} = p^2$$
$$p_{20} = (1 - p)^2$$
$$p_{21} = 2p(1 - p)$$
$$p_{22} = p^2$$

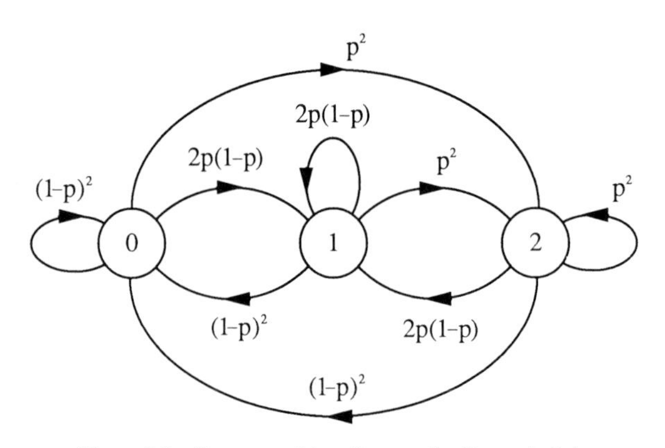


*Figure 3.1 State transition diagram for Example 3.1*

The corresponding state transition diagram is shown in Fig. 3.1 and should be self explanatory.

It would be nice if we had some means of determining performance measures, such as throughput and delay, for models like this. Intuitively, for the system in Example 3.1, the throughput is just the fraction of the time the channel spends in state 1, the transmit state. Alternatively, this may be thought of as the probability of finding the Markov chain in state 1 if it is observed at any random value of the time index. Such questions can be answered by finding the *equilibrium distribution* of a Markov chain, and we shall now address this problem.

### 3.3.2 Equilibrium distribution

First consider the transitions a Markov chain can make in two steps rather than one. The probability of moving from state $i$ to state $j$ in two steps is

$$p_{ij}^{(2)} = P(x_{n+2} = j|x_n = i), \qquad i, j \in \mathbf{X},\ n = 0, 1, 2, \ldots$$

Since in order to move from state $i$ to state $j$ in two steps, the chain must pass through some state $k$ after one step, then

$$p_{ij}^{(2)} = \sum_{k \in \mathbf{X}} P(x_{n+1} = k, x_{n+2} = j|x_n = i)$$

$$= \sum_{k \in \mathbf{X}} P(x_{n+2} = j|x_n = i, x_{n+1} = k)P(x_{n+1} = k|x_n = i) \tag{3.33}$$

Now, using the Markov property, equation (3.33) becomes

$$p_{ij}^{(2)} = \sum_{k \in \mathbf{X}} P(x_{n+2} = j|x_{n+1} = k)P(x_{n+1} = k|x_n = i)$$

$$= \sum_{k \in \mathbf{X}} p_{ik}p_{kj} \tag{3.34}$$

The right hand side of equation (3.34) is thus equal to the $ij$th element of $P^2$, the square of the transition probability matrix.

An inductive extension of the above leads to the result that the *m-step transition probabilities* of a Markov chain, defined by

$$p_{ij}^{(m)} = P(x_{n+m} = j|x_n = i), \qquad m = 0, 1, 2, \ldots \tag{3.35}$$

are the elements of the $m$th power of the transition probability matrix, $P^m$.

Now consider a Markov chain $\{x_n; n = 0, 1, 2, \ldots\}$ that is observed at some random value of the time index. When the moment of observation is infinitely removed from the starting point, the probability $\Pi_j$ of finding the chain in state $j$ becomes independent of the initial state $x_0$. That is

$$\Pi_j = \lim_{n\to\infty} P(x_n = j|x_0 = i)$$

$$= \lim_{n\to\infty} p_{ij}^{(n)}, \qquad j \in \mathbf{X} \tag{3.36}$$

When these limiting probabilities exist such that

$$\sum_{j\in\mathbf{X}} \Pi_j = 1 \tag{3.37}$$

they are called the equilibrium distribution of the Markov chain. Other names to describe this that are used in the literature are *steady-state distribution, long-term distribution*, and *invariant distribution.*

When a Markov chain has its equilibrium distribution, it is also said to be *stationary*. The motivation here is that if the equilibrium distribution is chosen as the initial distribution at $n = 0$, then the distribution will remain the same at $n = 1, 2, 3, \ldots, \infty$. That is, it is invariant.

Before discussing how to find the equilibrium distribution and the conditions under which it exists, we require some more definitions.

Let $v_{ij}$ be the probability that a Markov chain will eventually visit state $j$ having been in state $i$. That is

$$v_{ij} = P(x_n = j \textit{ for some } n|x_0 = i), \qquad i, j \in \mathbf{X} \tag{3.38}$$

If $v_{ij} \neq 0$, then state $j$ is *reachable* from state $i$. This also means that there must exist a path on the state transition diagram leading from state $i$ to state $j$ (it is implicit that each transition in this path has a non-zero probability).

A set of states $\mathbf{Y}$ is said to be *closed* if no state $j \in \mathbf{Y}^c$ is reachable from any state $i \in \mathbf{Y}$ ($\mathbf{Y}^c$ is the set of all states in $\mathbf{X}$ that are not in $\mathbf{Y}$). Clearly, $\mathbf{X}$ is closed. If $\mathbf{X}$ is the smallest set of states that is closed then the corresponding Markov chain in $\mathbf{X}$ is *irreducible.* Thus a Markov chain $\{x_n; n = 0, 1, 2, \ldots\}$ $x_n \in \mathbf{X}$, is irreducible on $\mathbf{X}$ if, and only if, every state $i \in \mathbf{X}$ is reachable from every other state $j \in \mathbf{X}$.

If $v_{ij} = 1$ for some $i \in \mathbf{X}$, then state $i$ is called *recurrent*. This means that having once visited a recurrent state $i$, the chain is certain to eventually visit it again. Thus if $i$ is a recurrent state, the Markov chain must either visit it not at all, or infinitely many times, since no visit to $i$ can be the last one.

If states are not recurrent they are called *transient*. Let $j$ be a transient state for some Markov chain, with probability, $v_{jj}$ that, after visiting $j$, the chain will eventually return to it, and let $K_j$ be a random variable representing the total number of visits to state $j$. Then $1 - v_{jj}$ is the probability that the chain visits state $j$ for the last time. Clearly, the total number of visits to state $j$ must be geometrically distributed with parameter

$1 - v_{jj}$. That is

$$P(K_j = k) = v_{jj}^{k-1}(1 - v_{jj}), \qquad k = 1, 2, \ldots \tag{3.39}$$

The mean number of visits is thus

$$E[K_j] = \frac{1}{1 - v_{jj}} \tag{3.40}$$

We therefore see that the total number of visits to state $j$ is finite if, and only if, $v_{jj} \neq 1$. Thus the number of visits to a transient state must be finite with probability 1. A consequence of this is that the equilibrium probability of a transient state must be zero.

Another repercussion of the transience and recurrence properties of the states is that if a Markov chain is irreducible, then either all its states are transient or all are recurrent. This can be shown by the following argument. If a state $i$ is recurrent, and $j$ is a transient state such that $i$ and $j$ are both members of the state space of an irreducible Markov chain, then $j$ must be reachable from $i$. Then if the chain visits state $j$, a return to $j$ is not certain. Then a visit to state $i$ is also not certain, since if it was, then the chain would keep visiting $i$ and eventually find its way back to $j$ (recall that $j$ is reachable from $i$). But this implies that it is possible for the chain to move from $i$ to $j$ and not return to $i$, which contradicts $i$ being recurrent. Thus states $i$ and $j$ must either both be recurrent or both transient.

It is usual to refer to Markov chains that have transient or recurrent states as *transient Markov chains* or *recurrent Markov chains*, respectively.

Now consider the evolution of an irreducible, recurrent Markov chain $\mathbf{x} = \{x_n; n = 0, 1, 2, \ldots\}$. Every state is visited by $\mathbf{x}$ infinitely many times with probability 1, irrespective of $x_0$. Let $\{n_i^j, i = 1, 2, \ldots\}$ be the times of consecutive visits to state $j$, and let $m_j$ be the expected interval between these visits. Then

$$m_j = E[n_{k+1}^j - n_k^j]; \qquad k = 1, 2, \ldots \tag{3.41}$$

Thus, since on average one out of every $m_j$ observations finds $\mathbf{x}$ in state $j$, the fraction of time spent in state $j$ by the Markov chain is $1/m_j$. Our intuition tells us that this fraction should equal the equilibrium probability, $\Pi_j$, of finding the chain in state $j$. This turns out to be true provided that the chain does not visit states periodically on a regular basis.

A state $j$ is called *periodic* with period $d > 1$ if the consecutive returns to $j$ can occur only at multiples of $d$ steps. More formally

$$P(x_{n+kd} = j \textit{ for some } k \geq 1 | x_n = j) = 1 \tag{3.42}$$

If no integer $d > 1$ exists that satisfies equation (3.42), then state $j$ is *aperiodic*. In a similar way that all states of an irreducible Markov chain are either transient or recurrent, it can be shown that all states of an irreducible Markov chain are either periodic with the same period, or all states are aperiodic. The corresponding Markov chains are called *periodic Markov chains* or *aperiodic Markov chains*, respectively.

The following result is now stated without proof. Further details can be found in [WOLF 89].

*First limiting result:* If $\mathbf{x} = \{x_n, n = 0, 1, 2, \ldots\}$ is an irreducible, aperiodic and recurrent Markov chain, then the limiting probabilities defined in equation (3.36) exist and are given by

$$\Pi_j = \frac{1}{m_j}; \qquad j \in \mathbf{X} \tag{3.43}$$

Although the Markov chain is specified as recurrent, if the intervals between visits to a given state are infinitely large, such states are called *recurrent null.* Clearly, such states will have zero equilibrium probabilities. States with non-zero equilibrium probabilities (and hence finite return times) are called *recurrent non-null* (or *positive recurrent*). Again it can be shown that all states of an irreducible Markov chain must be either recurrent non-null, or all states must be recurrent null. The corresponding Markov chains are also called *recurrent non-null Markov chains* or *recurrent null Markov chains.*

Finally, the following result tells us how the equilibrium distribution of a Markov chain can be determined, and the conditions under which it is unique.

*Steady-state theorem:* An irreducible and aperiodic Markov chain $\mathbf{x} = \{x_n = x; n = 0, 1, 2, \ldots\}$, $x \in \mathbf{X}$, with transition probability matrix $P = [p_{ij}; i, j \in \mathbf{X}]$, is recurrent non-null if, and only if, the system of equations

$$\Pi_j = \sum_{i \in \mathbf{X}} \Pi_i p_{ij}, \qquad j \in \mathbf{X} \tag{3.44}$$

$$\sum_{j \in \mathbf{X}} \Pi_j = 1 \tag{3.45}$$

has a solution. That solution is then unique, and is the equilibrium distribution of $\mathbf{x}$. Equations (3.44) are called the *balance equations,* and equation (3.45) is called the *normalising equation* of the Markov chain $\mathbf{x}$.

Equations (3.44) are called balance equations because they equate the fraction of all transitions out of state $j$ with the fraction of all transitions entering state $j$. To see this, consider the following equation:

$$\sum_{\substack{i \in \mathbf{X} \\ i \neq j}} \Pi_j p_{ji} = \sum_{\substack{i \in \mathbf{X} \\ i \neq j}} \Pi_i p_{ij} \tag{3.46}$$

A little thought will show that the left hand side of equation (3.46) represents the fraction of all transitions out of state $j$ to all other states $i \neq j$. Likewise, the right hand side represents the fraction of all transitions into state $j$

from all other states $i \neq j$. With the Markov chain in equilibrium, equation (3.46) must hold for all states $j \in \mathbf{X}$, and is simply an expression of the principle of flow balance which applies to any system that has attained a steady-state.

Now adding $\Pi_j p_{jj}$ to each side of equation (3.46), and since $\Pi_j$ on the left hand side is independent of the summation index, we have

$$\Pi_j p_{jj} + \Pi_j \sum_{\substack{i \in \mathbf{X} \\ i \neq j}} p_{ji} = \Pi_j p_{jj} + \sum_{\substack{i \in \mathbf{X} \\ i \neq j}} \Pi_i p_{ij}$$

This is equivalent to

$$\Pi_j \sum_{i \in \mathbf{X}} p_{ji} = \sum_{i \in \mathbf{X}} \Pi_i p_{ij}$$

Thus, using equation (3.32) we have

$$\Pi_j = \sum_{i \in \mathbf{X}} \Pi_i p_{ij}$$

Since this holds for all $j \in \mathbf{X}$, then we have equations (3.44), which thus simply equate the fraction of transitions into and out of all states $j \in \mathbf{X}$.

The sufficiency of the steady-state theorem can be shown by letting $x_0$ be chosen at random with the distribution $\Pi_j$, $j = 0, 1, 2, \ldots$. Then since the balance equations are satisfied, the same distribution must apply at all subsequent observation instants, no matter how far removed from the start. This implies that the Markov chain can be neither transient nor recurrent-null, since in either case all the equilibrium probabilities would be zero, contradicting equation (3.45).

Equation (3.43) means that the Markov chain cannot have more than one equilibrium distribution, thus from this and the sufficiency of the steady state theorem, then a solution to equations (3.44) and (3.45), if it exists, must be unique.

We thus arrive at the important result that the equilibrium analysis of a system modelled by an aperiodic, irreducible Markov chain reduces to the solution of the corresponding balance equations in conjunction with the normalising equation. This can range from a trivial task to an extremely difficult one, depending on the size of the state space and on the symmetry or other exploitable structure exhibited by the one-step transition probability matrix.

A Markov chain with a finite state-space, $\mathbf{x} = \{x_n = x; n = 0, 1, 2, \ldots\}$, $x \in \mathbf{X} = \{0, 1, 2, \ldots, H\}$, is always recurrent non-null and so equations (3.44) and (3.45) must have a unique solution. That this must be so is easy to see in that if $\mathbf{x}$ is transient or recurrent null, then the equilibrium probabilities $\Pi_j$ of all states $j \in \mathbf{X}$ are zero. Using equation (3.36), and summing over the (finite) state space

$$\sum_{j=0}^{H} \Pi_j = \sum_{j=0}^{H} \lim_{n\to\infty} P(x_n = j|x_0 = i) \tag{3.47}$$

Now if **x** is transient or recurrent null

$$\sum_{j=0}^{H} \Pi_j = 0 \tag{3.48}$$

However, since at any time $n = 0, 1, 2, \ldots$, **x** must be in some state with probability 1, we have the identity

$$\sum_{j=0}^{H} P(x_n = j) = 1 \tag{3.49}$$

which must hold for all $n$, regardless of the initial state. This therefore implies that

$$\sum_{j=0}^{H} \lim_{n\to\infty} P(x_n = j|x_0 = i) = 1 \tag{3.50}$$

Clearly, we have a contradiction that can only be resolved by the deduction that **x** must be recurrent non-null.

Since many practical systems can be modelled by Markov chains that have a finite state space, this is an important result in that provided the Markov chains are irreducible and aperiodic, then a unique equilibrium distribution always exists, and can be obtained as a solution to the balance equations and the normalising equation.

An interesting point in this context is that when the state-space is finite, one of the balance equations is always linearly dependent on the others under the constraint of the normalising equation. This is not a problem however, since the linear dependence can be removed in the usual way by dropping one of the balance equations and solving the remaining system of $H + 1$ equations (the $H$ remaining balance equations plus the normalising equation) simultaneously to obtain the $H + 1$ unknowns, $\Pi_0, \Pi_1, \ldots, \Pi_H$.

As a simple illustration of finding the equilibrium distribution, we shall consider again the Markov chain defined in Example 3.1.

*Example 3.2: Equilibrium distribution*

Determine the equilibrium distribution of the Markov chain defined in Example 3.1 as a function of the parameter $p$. Use the equilibrium distribution to find the channel throughput normalised to message transmission time.

*Solution:* For this system the balance equations are

$$\Pi_0 = \Pi_0 p_{00} + \Pi_1 p_{10} + \Pi_2 p_{20}$$

$$\Pi_1 = \Pi_0 p_{01} + \Pi_1 p_{11} + \Pi_2 p_{21}$$

$$\Pi_2 = \Pi_0 p_{02} + \Pi_1 p_{12} + \Pi_2 p_{22}$$

Substitution for the transition probabilities from Example 3.1, and using the normalising equation, then

$$\Pi_0 = (1 - p)^2$$

$$\Pi_1 = 2p(1 - p)$$

$$\Pi_2 = p^2$$

The equilibrium distribution is thus

$$\Pi = \left\{ \Pi_K = \binom{2}{2-K} p^K (1-p)^{2-K}; \qquad K \in \{0, 1, 2\} \right\}$$

The throughput, $S$, is just the fraction of time the channel spends transmitting packets. That is

$$S = \Pi_1 = 2p(1 - p)$$

Other performance measures, such as mean message delay, might also be obtained from the equilibrium distribution. These issues will be considered later.

### 3.3.3 Reversible chains

It can be observed that if a Markov chain is stationary, then it also behaves as a Markov chain when its evolution is observed in *reversed time*. This is shown by the following, which is a result that leads to a verification lemma (Kelly's Lemma) [KELL 79] that enables us to guess and verify the equilibrium distribution of certain Markov chains.

*Time reversal theorem:* Let $\mathbf{x} = \{x_n = x, n = 0, 1, 2, \ldots\}$ be a stationary Markov chain with initial distribution $\Pi_0 = \Pi$ and with transition matrix $P$. This will be written as $\mathbf{x} = (\Pi, P)$. Then for any integer $T \geq 0$, $\{x_{T-n}, n = 0, 1, \ldots, T\}$ has the *same law* as a Markov chain $(\Pi, \tilde{P})$ on $\{0, 1, \ldots, T\}$, where $\tilde{P} = [\tilde{p}_{ij}; i, j \in \mathbf{X}]$ with

$$\Pi_i \tilde{p}_{ij} = \Pi_j p_{ji}; \qquad i, j \in \mathbf{X} \tag{3.51}$$

In particular, the Markov chain has the same law when reversed in time if and only if $\tilde{P} = P$; that is, if and only if

$$\Pi_i p_{ij} = \Pi_j p_{ji}; \qquad i, j \in \mathbf{X} \tag{3.52}$$

In this latter case, $\mathbf{x}$ is said to be *time-reversible*. This somewhat standard terminology thus means that a time-reversed Markov chain is not necessarily time-reversible!.

To show that equation (3.51) is correct, consider $n = 0, 1, \ldots, T$, and $\tilde{x}_n = x_{T-n}$; then

$$\begin{aligned} P\{\tilde{x}_0 = i_0, \ldots, \tilde{x}_T = i_T\} &= P\{x_0 = i_T, \ldots, x_T = i_0\} \\ &= \Pi_{i_T} p_{i_T i_{T-1}} p_{i_{T-1} i_{T-2}} \cdots p_{i_1 i_0} \\ &= \tilde{p}_{i_{T-1} i_T} \Pi_{i_{T-1}} p_{i_{T-1} i_{T-2}} \cdots p_{i_1 i_0} \\ &= \ldots \\ &= \tilde{p}_{i_{T-1} i_T} p_{i_{T-2} i_{T-1}} \cdots \tilde{p}_{i_0 i_1} \Pi_{i_0} \end{aligned}$$

where the last equalities follow directly from equation (3.51). This therefore shows that $\tilde{x}_n$ behaves as the Markov chain $(\Pi, \tilde{P})$.

This result leads to the following lemma:

*Kelly's Lemma:* Assume that $\Pi$ is a distribution, and $P$, $P'$ are two transition matrices on $\mathbf{X}$ such that

$$\Pi_i p_{ij} = \Pi_j p'_{ji}; \qquad i, j \in \mathbf{X} \tag{3.53}$$

Then $\Pi P = \Pi$, and $P' = \tilde{P}$, the transition matrix of $(\Pi, P)$ reversed in time.

As an example of a time-reversible Markov chain, consider the following.

*Example 3.3: Communication buffer as a time-reversible Markov chain*

Messages arrive at a communication buffer (that has infinite capacity) with probability $\alpha$ per unit time according to a Bernoulli process (zero or one arrival per unit time), and are queued on a first-come first-served basis. The message at the head of the queue (if any) is transmitted and departs from the buffer with probability $\beta$ per unit time, the arrivals and departures being mutually exclusive such that $\alpha + \beta = 1$. The number of messages in the buffer is represented by a Markov chain $\mathbf{x} = \{x_n, n = 0, 1, 2, \ldots\}$ with transition matrix $[p_{ij}; i, j \in \mathbf{X} = \{0, 1, 2, \ldots\}]$. The task is to find the equilibrium distribution of $\mathbf{x}$, and show that the Markov chain is time-reversible, assuming $\alpha\beta > 0$.

*Solution:* From the problem specification, the transition probabilities are defined by

$$p_{ij} = \begin{cases} 0; & j < i - 1, j > i + 1, j = i \\ \alpha; & j = i + 1 \\ \beta; & j = i - 1, i \neq 0 \\ 1; & j = 1, i = 0 \end{cases}$$

The balance equations are

$$\Pi_0 = \Pi_1 \beta$$

$$\Pi_1 = \Pi_0 + \Pi_2 \beta$$

$$\Pi_K = \Pi_{K-1}\alpha + \Pi_{K+1}\beta; \qquad K = 2, 3$$

These solve recursively to give

$$\Pi_1 = \Pi_0/\beta$$

$$\Pi_2 = \Pi_0(\alpha/\beta)/\beta$$

$$\Pi_3 = \Pi_0(\alpha/\beta)^2/\beta$$

$$\dots$$

$$\Pi_K = \Pi_0(\alpha/\beta)^{K-1}/\beta; \qquad K = 1, 2, \dots$$

Normalising the distribution, then

$$\sum_{K=0}^{\infty} \Pi_K = \Pi_0 \left[ 1 + \sum_{K=1}^{\infty} (\alpha/\beta)^{K-1}/\beta \right] = 1$$

This can be done only if $\alpha < \beta$, in which case the infinite sum on the right converges. In this case

$$\Pi_0 = (\beta - \alpha)/2\beta$$

and

$$\Pi_K = (\alpha/\beta)^{K-1}(\beta - \alpha)/2\beta^2; \qquad K = 1, 2, \dots$$

It turns out that all states are transient if $\alpha > \beta$, and all states are null if $\alpha = \beta$.

To show that this Markov chain is time reversible, we need only verify equation (3.52) for positive elements of the transition matrix $P$. Then

$$\Pi_0 = \Pi_1 \beta$$

$$\Pi_K \alpha = \Pi_{K+1}\beta, \qquad K = 1, 2, \dots$$

The solution obtained for $\Pi$ thus satisfies equation (3.52), and so the chain is reversible provided that $\alpha < \beta$.

The example can be generalised to some extent by allowing the probabilities $\alpha$ and $\beta$ to depend on the state index $K$. Let $p_{i,i+1} = \alpha_i$, $p_{i,i-1} = \beta_i = 1 - \alpha_i$, where $\alpha_i \beta_i > 0$, $i = 1, 2, \dots$. Then

$$\Pi_0 = \beta_1 \Pi_1$$

$$\Pi_K \alpha_K = \Pi_{K+1}\beta_K, \qquad K = 1, 2, \dots$$

Hence, this chain is reversible under the same normalisation conditions for it to be recurrent non-null.

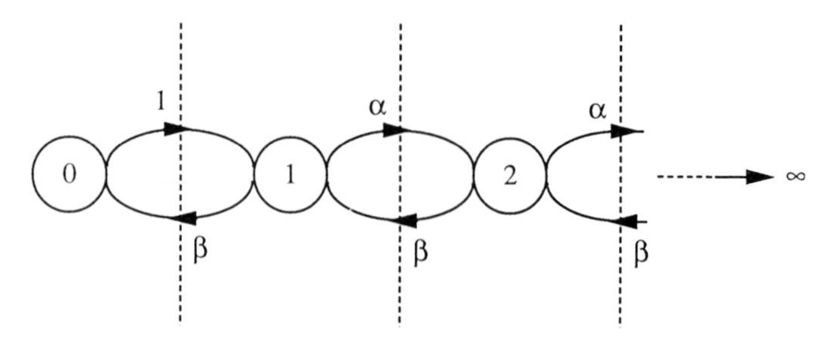


*Figure 3.2 State transition diagram for Example 3.3*

The key to this example is that because the one step transitions are between adjacent states, equation (3.52), in effect, equates the *transition rates* (transitions per unit time) across the boundary between sets of the form $\{0, \ldots, i\}$ and $\{i+1, i+2, \ldots\}$.

These boundaries are indicated by the dotted lines in Fig. 3.2, which shows the state transition diagram for Example 3.3. Because state changes are either $+1$ or $-1$, the chain can return to a state only after an even number of transitions; that is, the chain is periodic with period $d = 2$. By allowing $p_{ii} > 0$ for some $i$, the chain would be aperiodic, yet it would still be reversible. However, if we let $p_{ij} > 0$ for some pair of states in a recurrent non-null chain with period $d \geq 3$, then clearly $\Pi_i p_{ij} > 0$. Then we must have $\Pi_j p_{ji} = 0$, otherwise $p_{ji} > 0$, and the chain would have period $d \leq 2$. Hence, a periodic chain with period $d \geq 3$ *cannot be reversible.* However, reversible chains with period $d \leq 2$ are important special cases, as shown by Example 3.3, which is a simple example of a *discrete-time queue.*

The concept of equating transition rates across boundaries can be generalised in that for any Markov chain that satisfies the balance equations (3.44), then for all $\mathbf{A} \subset \mathbf{X}$

$$\sum_{i \in \mathbf{A}} \sum_{j \in \mathbf{A}^c} \Pi_i p_{ij} = \sum_{j \in \mathbf{A}^c} \sum_{i \in \mathbf{A}} \Pi_j p_{ji} \tag{3.54}$$

with $\mathbf{A}^c = \mathbf{X} - \mathbf{A}$. To show this, let

$$F(\mathbf{A}, \mathbf{B}) = \sum_{i \in \mathbf{A}} \sum_{j \in \mathbf{B}} \Pi_i p_{ij}$$

Then we have

$$\begin{aligned} F(\mathbf{A}, \mathbf{X}) &= \sum_{i \in \mathbf{A}} \sum_{j \in \mathbf{X}} \Pi_i p_{ij} \\ &= \sum_{i \in \mathbf{A}} \Pi_i \sum_{j \in \mathbf{X}} p_{ij} \\ &= \sum_{i \in \mathbf{A}} \Pi_i \end{aligned}$$

This latter equality is because $P$ is a stochastic matrix, and so each row $i$ must sum to 1.

Also we have

$$F(\mathbf{X}, \mathbf{A}) = \sum_{j \in \mathbf{X}} \sum_{i \in \mathbf{A}} \Pi_j p_{ji}$$

$$= \sum_{i \in \mathbf{A}} \sum_{j \in \mathbf{X}} \Pi_j p_{ji}$$

$$= \sum_{i \in \mathbf{A}} \Pi_i$$

This latter equality is because the Markov chain solves the balance equations (3.44).

Finally, we thus have

$$F(\mathbf{A}, \mathbf{A}^c) = F(\mathbf{A}, \mathbf{X}) - F(\mathbf{A}, \mathbf{A})$$

$$= F(\mathbf{X}, \mathbf{A}) - F(\mathbf{A}, \mathbf{A})$$

$$= F(\mathbf{A}^c, \mathbf{A})$$

which is the desired result. This result therefore says that provided a Markov chain satisfies the balance equations, then the transition rates *across arbitrary boundaries* must be the same.

The concept of time reversibility can be extended to Markov chains whose state transition diagram takes the form of a *tree*. More formally, we say that the state transition diagram of a Markov chain $(\Pi, P)$ is a tree if $p_{ij} > 0$ whenever $p_{ji} > 0$, and if the set $\mathbf{G} = \{(i, j) \in \mathbf{X}^2 | p_{ij} > 0\}$ is a tree, with the obvious meaning of the term. The Markov chain $(\Pi, P)$ is then time-reversible when it is stationary.

To check that such a Markov chain satisfies equation (3.52), we note that to any pair $(i, j) \in \mathbf{G}$ corresponds a set $\mathbf{A}$ such that $i \in \mathbf{A}$, $j \in \mathbf{A}^c$, and $(i, j)$ is the only such pair $(k, l)$ in $\mathbf{A} \times \mathbf{A}^c$ such that $p_{kl} > 0$. This is illustrated in Fig. 3.3. Equation (3.52) then follows directly from equation (3.54).

To establish that a chain is reversible from equation (3.52), we must find the equilibrium probability distribution $\Pi$. An obvious question is to ask if reversibility can be established without doing this. First note that when $P = \tilde{P}$, paths in the forward and reverse direction have the same probability. For example,

$$p_{ij} p_{jk} p_{kl} = \tilde{p}_{ij} \tilde{p}_{jk} \tilde{p}_{kl}$$

$$= (p_{ji} \Pi_j / \Pi_i)(p_{kj} \Pi_k / \Pi_j)(p_{lk} \Pi_l / \Pi_k)$$

or

$$\Pi_i p_{ij} p_{jk} p_{kl} = \Pi_l p_{lk} p_{kj} p_{ji} \tag{3.55}$$

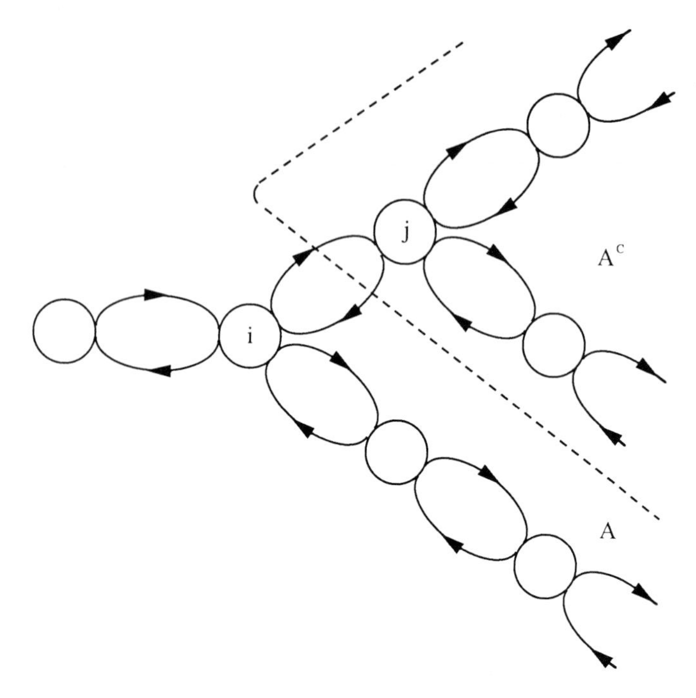


*Figure 3.3 Time-reversible Markov chain as a tree*

When $l = i$ in equation (3.55), we can cancel $\Pi_i$ to give

$$p_{ij}p_{jk}p_{kl} = p_{ik}p_{kj}p_{ji} \tag{3.56}$$

When the initial and terminal states of a path are the same, the path is called a *loop*. This leads to the following result.

*Kolmogorov's Criterion:* A stationary Markov chain is reversible if and only if the probability of traversing any loop is equal to the probability of traversing the same loop in the reverse direction.

The necessity of this result follows from the argument leading to equation (3.56). To prove sufficiency, the criterion implies that for any pair of states $i \neq j$ and any $n \geq 1$

$$p_{ij}p_{ji}^{(n)} = p_{ij}^{(n)}p_{ji} \tag{3.57}$$

To see this, note that there is a one-to-one correspondence between loops with probability included on the left hand side of equation (3.57) and loops traversed in the reverse direction with probability on the right. Then from equation (3.57)

$$\left(\sum_{S=1}^{n} p_{ji}^{(S)}/n\right)p_{ij} = \left(\sum_{S=1}^{n} p_{ij}^{(S)}/n\right)p_{ji}, \qquad n = 1, 2, \ldots \tag{3.58}$$

Letting $n \to \infty$ in equation (3.58) gives equation (3.52).

Kolmogorov's criterion thus enables reversibility to be established directly from the transition probabilities (provided that the Markov chain is stationary).

Although Markov chains that are time-reversible are obviously special cases, the concept of time-reversal is an important one, and will be used later for proving results on discrete-time queues. The topic has been studied in some detail in the context of continuous time systems, the classic text being [KELL 79].

## 3.4 MARKOV CHAIN MODELS

One way of modelling the performance of a communication or computer network is to decide on a sufficient state description and then model the system as a Markov chain. Since many physical systems such as communication networks operate on a time-slotted basis, these can be conveniently modelled by discrete-time Markov chains by specifying a one-to-one correspondence between a time-slot in the physical system and unit time in the model. Performance measures for the system can then be extracted from the equilibrium distribution of the Markov chain. Although the equilibrium distribution can, of course, be obtained by a direct solution of the balance equations and normalisation, this is a somewhat brute-force modelling approach in many situations. In principle, any Markov chain that is irreducible, aperiodic, and recurrent non-null can be solved by this direct approach. In practice, whether or not such an approach is feasible depends on a number of factors such as the size of the state space, the structure of the transition probability matrix, and the degree of difficulty in calculating this latter.

In many cases a direct solution of the balance equations is only possible if the model is first simplified. Another approach is to use an exact model, but an approximate solution technique. This latter approach will generally be favoured, although the direct solution technique is obviously the method to use if the model is simple enough. It is illustrated in the next example, which is a communication network that uses a variant of the well known slotted Aloha protocol [TOBA 80B].

*Example 3.4: Unbuffered slotted Aloha with delayed first transmission.*

A model for this system is shown in Fig. 3.4. The network has a fixed population of $N$ users, and time is divided into slots such that in any given slot a user can be at one of the two nodes in Fig. 3.4 , I (idle) or

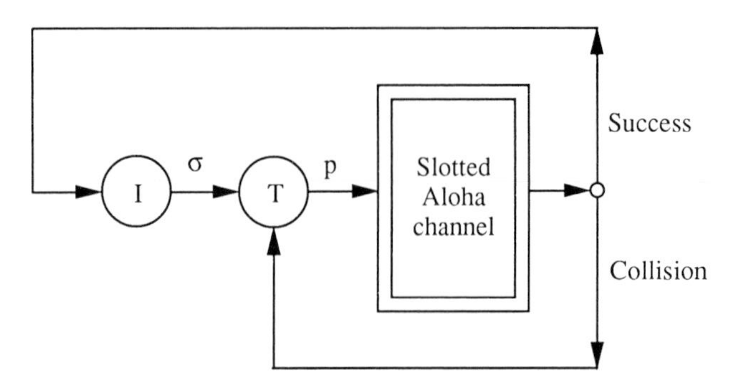


*Figure 3.4 Markovian model for unbuffered slotted Aloha with delayed first transmission*

T (transmit). Users at node I have no messages for transmission and each user generates new messages with probability $\sigma$ per slot, the message generation assumed to occur at the start of a slot. Users at node I that generate a message jump to node T at the end of the slot. Users at node T each have exactly one message waiting for transmission, and they attempt to transmit this with probability $p$ at the end of each slot, until it is successfully transmitted. If two or more users at node $T$ attempt transmission at the end of a slot, the messages collide and the corresponding users circle the collision looop and immediately return to node T. If exactly one user attempts a transmission from node T at the end of a slot, the transmission is successful, and the corresponding user jumps to node I.

The assumptions made are that users are statistically identical, messages are of fixed length equal to one slot, the channel has zero propagation delay and is free of transmission errors, and users at node T are not allowed to generate a new message.

The objective is to calculate the mean throughput of the network by modelling this as a discrete-time Markov chain and directly solving the balance equations. Note that the assumptions made might not be realistic in many situations, but are necessary in order to make the model tractable by direct Markov analysis.

*Solution:* The first point to note is that if there are $x$ users at node T in any given slot, then there are $N - x$ users at node I. The behaviour of this system can thus be described by the Markov chain

$$\mathbf{x} = \{x_n = x; \qquad n = 0, 1, 2, \ldots\}$$

where for a given $N$, $x_n$ can take values $x \in \mathbf{X}_N$, where $\mathbf{X}_N$ is the state space defined by

$$\mathbf{X}_N = \{0, 1, 2, \ldots, N\}$$

The state transition diagram indicating the possible state transitions is

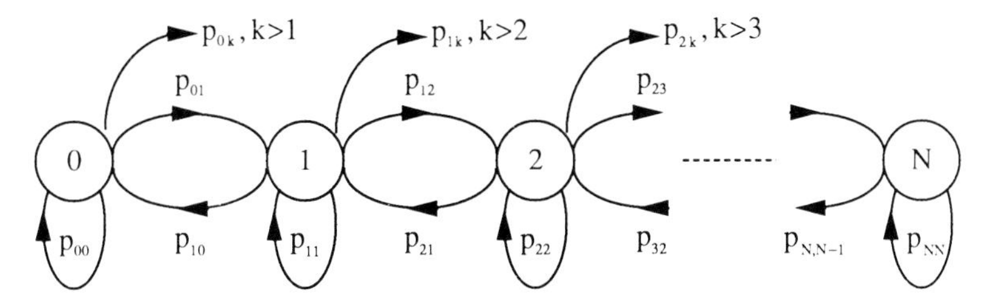


*Figure 3.5 Possible state transitions for the Markov chain of Example 3.4*

shown in Fig. 3.5. The one step transition probabilities are given by

$$p_{ij} = P(x_{n+1} = j | x_n = i)$$

These can be calculated as follows:
$j \leq i - 2$: Not more than one successful transmission from users at node T is possible in a slot, and so downward transitions of the two or more states are impossible. Thus:

$$p_{ij} = 0, j \leq i - 2 \tag{3.59}$$

$j = i - 1$: Downward transitions of exactly one state occur if the two following independent events occur:

1. There is exactly one successful transmission from users at node T. This event has probability $ip(1 - p)^{i-1}$.
2. There are no new message arrivals for users at node I. This event has probability $(1 - \sigma)^{N-i}$.

Thus:

$$p_{ij} = ip(1 - p)^{i-1}(1 - \sigma)^{N-i}, \qquad j = i - 1 \tag{3.60}$$

$j = i$: The state of the Markov chain will remain the same if the number of users leaving node T for node I balance the number of users leaving node I for node T. This can occur in two possible ways:

1. There are no message arrivals for users at node I, and there are zero or more than one transmission attempts from users at node T. That is, no user leaves node I, and either no user leaves node T, or users go around the collision loop in Fig. 3.4 and immediately return to node T. This joint event has probability

   $$(1 - \sigma)^{N-i}[1 - ip(1 - p)^{i-1}]$$

2. There is exactly one message arrival for users at node I, and one successful transmission from users at node T. That is, a single user jumping from node I to node T balances a single user jumping from node T to node I. This joint event has probability

   $$(N - i)\sigma(1 - \sigma)^{N-i-1}ip(1 - p)^{i-1}$$

Thus:

$$p_{ij} = (1-\sigma)^{N-i}[1 - ip(1-p)^{i-1}] + (N-i)\sigma(1-\sigma)^{N-i-1}ip(1-p)^{i-1} \tag{3.61}$$

$j > i$: This implies the state of the Markov chain jumps up by $j - i$. This can happen in two possible ways:

1. There are $j - i$ message arrivals for users at node I and there are zero or more than one transmission attempts from users at node T. That is, $j - i$ users jump from node I to node T, and either no users leave node T, or users that do leave go around the collision loop and immediately return. This joint event has probability

$$\binom{N-i}{j-i}\sigma^{j-i}(1-\sigma)^{N-j}[1 - ip(1-p)^{i-1}]$$

2. There are $j - i + 1$ message arrivals for users at node I, and there is a successful transmission from a user at node T. Thus $j - i + 1$ users jump from node I to node T, and one user jumps from node T to node I, giving a net increase of $j - i$ users at node T. This joint event has probability

$$\binom{N-i}{j-i+1}\sigma^{j-i+1}(1-\sigma)^{N-j-1}ip(1-p)^{i-1}$$

Thus:

$$p_{ij} = \binom{N-i}{j-i}\sigma^{j-i}(1-\sigma)^{N-j}[1 - ip(1-p)^{i-1}] + \binom{N-i}{j-i+1}\sigma^{j-i+1}(1-\sigma)^{N-j-1}ip(1-p)^{i-1} \tag{3.62}$$

This completes the calculation of the one-step state transition probabilities, which can now be evaluated explicitly for specified values of $\sigma$, $p$ and $N$. The balance equations

$$\Pi_j = \sum_{i \in \mathbf{X}_N} \Pi_i p_{ij}, \qquad j = 0, 1, \ldots, N$$

and the normalising equation

$$\sum_{i \in \mathbf{X}_N} \Pi_i = 1$$

can now be solved for $\Pi_j$, keeping in mind that because of the finite state space one of the balance equations must be discarded due to the linear dependence amongst them. The conditional throughput of the system, given that the Markov chain is at state $i$, is

$$S(i) = ip(1-p)^{i-1} \tag{3.63}$$

The mean throughput is now given by

$$S = \sum_{i \in \mathbf{X}_N} \Pi_i S(i) \tag{3.64}$$

Several observations can be made from this example. Firstly, even for such a simple system the calculation of the one-step state transition probabilities is surprisingly involved. It is easy to envisage these calculations becoming prohibitively complex for very complicated models. Secondly, the Markov chain was easy to specify because the state had a very simple interpretation (the number of users at node T) as a *scalar* quantity. In many cases the state description is a *vector*, thus giving a *multidimensional Markov chain*. For example, if we drop the assumption in Example 3.4 that the channel has zero propagation delay, then the system model would appear as in Fig. 3.6, which shows the same system with a delay of $R$ slots on the channel. That is, any user successfully transmitting at the end of a slot from node $T_0$ only arrives at node I after a delay of $R$ slots. Similarly, colliding users will only return to node $T_0$ after an $R$ slot delay.

If the nodes are labelled as in Fig. 3.4, then if $x_n^i$ is a random variable representing the number of users at node $T_i$, $i = 0, 1, 2, \ldots, 2R$, the state description of the system becomes the vector

$$x_n = (x_n^0, x_n^1, x_n^2, \ldots, x_n^{2R}) \tag{3.65}$$

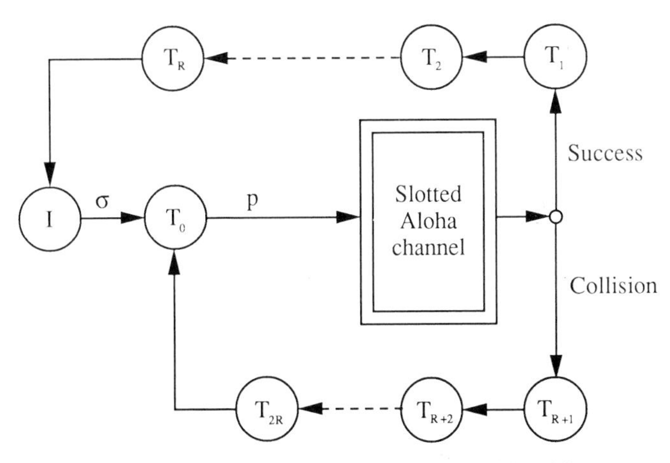


*Figure 3.6 Markovian model for unbuffered slotted Aloha with delayed first transmission and finite channel delay*

The number of users at node I in slot $n$, $x_n^I$, is then

$$x_n^I = N - \sum_{i=0}^{2R} x_n^i \tag{3.66}$$

Because the state space is finite, then, in principle, the states can be labelled to give a single dimensional Markov chain. In practice, unless the state space is not too large and the matrix, $P$, has an exploitable structure, then an exact Markov analysis becomes intractable.

In such cases, rather than simplify an exact model by making possibly unrealistic assumptions, it is usually better to solve an exact model using some form of approximation technique. The use of such methods will be considered in detail later.

## 3.5 EXERCISES

3.1 A communication buffer can store $J$ fixed length messages. New messages arrive at the buffer with probability $\alpha$ per unit time. Also, in any time unit the buffer can be completely purged, which happens with probability $\beta$. If both an arrival and a purging occur in the same time unit, the arrival is assumed to take place first. If the number of messages in the buffer at time $n$ is $\{x_n, j = 0, 1, 2, \ldots\}$, and the arrival and purging processes are independent of each other and of their past histories, find the equilibrium distribution of the Markov chain $\mathbf{x} = \{x_n = x, n = 0, 1, 2, \ldots\}$, $x \in \{0, 1, 2, \ldots, J\}$.

3.2 Repeat Exercise 3.1 for the case that the buffer is only purged in a time unit if the number of messages stored in the buffer is greater than or equal to some value $M$, $1 \leq M \leq J$.

3.3 The standard model for slotted Aloha appears as in Fig. E3.3.

A user at node I is idle and can generate a new message in a slot with probability $\sigma$, with the message generation assumed at the start of a slot.

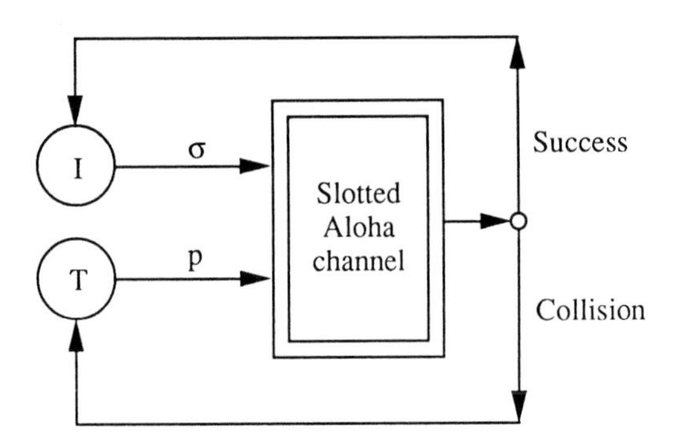


*Figure E3.3 Markovian model for slotted Aloha*

If a new message is generated, the corresponding user immediately attempts to transmit this in the current slot. A user at node T has a message waiting for retransmission, and will attempt to retransmit this with probability $p$ per slot. If exactly one user attempts a transmission/retransmission in a slot, this is successful, and the corresponding user jumps to node I at the end of the slot. If two or more users attempt a transmission/retransmission in a slot, a collision occurs and the users jump to node T at the end of the slot. If no transmissions/retransmissions are attempted in a slot, the users remain at their respective nodes. If $\{x_n, n = 0, 1, 2, \ldots\}$ is the number of users at node T in slot $n$, calculate the transition probabilities of the Markov chain $\mathbf{x} = \{x_n = x, n = 0, 1, 2, \ldots\}$, $x \in \{0, 1, 2, \ldots, N\}$, where $N$ is the number of users in the system. Find an expression for the conditional throughput of the network, given that the Markov chain is at state $x$, as a function of $\sigma$, $p$ and $N$.

# 4. DISCRETE-TIME QUEUES

*Chapter Objectives: To introduce some basic, discrete-time queueing systems, and methods that can be used to analyse them.*

Many of the problems in the performance modelling of communication or computer networks can be approached using the theory of queues. This theory has its origins in 1917 when the Danish engineer A. K. Erlang probabilistically analysed the behaviour of some queueing models that reflected certain performance features of telephone exchanges. Since then the theory has been gradually developed, in particular over the last twenty years, when much of the application has been directed towards digital computer and communications systems performance modelling.

Most of the vast literature associated with queueing systems however is devoted largely to continuous-time systems, whereas the developments in the practical world of computers and communications are becoming more and more digital, or discrete-time, in nature. In many situations, digital systems would seem to lend themselves far more readily to a discrete-time approach to modelling and analysis; so why the relative dearth of literature concerning discrete-time queueing systems when compared to that available for their continuous-time counterparts? This appears to stem from the fact that continuous-time queues can be significantly less complex to model than equivalent discrete-time ones. In continuous-time, only single events, such as packet arrivals or departures, can possibly occur at any given time instant. In discrete-time systems, the basic time unit, usually called a slot, is of finite size, and so the possibility of multiple events taking place in this time unit must be allowed for. In many situations however, the simplifications that can be effected by modelling and analysing a digital system in discrete-time compared with a continuous-time modelling can outweigh the additional complexities inherent in a discrete-time approach. Perhaps one of the earliest references on discrete-time queueing is [MEIS 58].

Section 4.1 deals with performance measures, including a simple heuristic derivation of Little's result. This latter is of fundamental importance in queueing theory. Section 4.2 covers some basic discrete-time queueing conventions that are used. Section 4.3 analysis the discrete-time $M/M/1$ queue as a Markov chain, and Section 4.4 extends these results to the finite capacity, discrete-time $M/M/1/J$ queue. Section 4.5, in turn, extends the discrete-time $M/M/1$ result to account for multiple (batch) arrivals in a slot, namely, the discrete-time $M^{a_n}/M/1$ queue, where $a_n$ is the number of arrivals in slot $n = 0, 1, 2, \ldots$. This is done by means of the generating function approach. Section 4.6 discusses the $M^{a_n}/M^{d_n}/\infty$ queue, which considers both multiple arrivals and departures in a slot. The number of servers in this queue is infinite. Section 4.7 introduces a class of discrete-time queues called $S$-queues; the $M^{a_n}/M^{d_n}/\infty$ queue, where $\{a_n\}$ is a sequence of i.i.d. Poisson random variables, is seen to be a special case of an $S$-queue. $S$-queues are later used to construct networks of queues and models for communication networks.

## 4.1 PERFORMANCE MEASURES AND LITTLE'S RESULT

### 4.1.1 Performance measures

We shall be concerned with two main performance measures: the waiting time and the throughput. These will be given the symbols $W$ and $S$, respectively, to be consistent with the notation introduced in Section 1.4 which discussed the performance measures of a multiple access protocol.

*Waiting time*, $W$, of a queueing system is defined as the time interval from when an arbitrary customer enters the system to the time the customer leaves, including the time spent in service. A variant of this is the *queueing time*, $Q$, defined as the time interval from when an arbitrary customer enters the system to the time the customer leaves, excluding the time spent in service.

*Throughput*, $S$, of a queueing system is defined as the number of customers passing through the system per unit time. *Normalised throughput*, which will be given the same symbol, $S$, is defined as the number of customers passing through the system per mean customer service time.

A secondary performance measure is the *probability of blocking*, $P_B$, which is defined as the fraction of customers that arrive to find no waiting room available to accommodate them. Clearly, this measure is only meaningful when the waiting room in the system is finite.

In discrete-time queueing systems the units of throughput are customers per slot, the units of normalised throughput are customers per mean customer service time, and waiting time has units of slots, or perhaps units of mean customer service times.

### 4.1.2 Little's result

The single most important performance measure in any queueing system is the mean waiting time. Little's result specifies a relation between the mean waiting time, $W$, the mean arrival rate of customers into the system, $\lambda$, and the mean number of customers stored in the system, $L$, as $L = \lambda W$. Note that the same symbols are used here for mean values as for arbitrary values. This policy will generally be adopted where no ambiguity arises.

Since $L$ is often easy to calculate, and $\lambda$ is usually specified, then Little's result gives a simple way of calculating the mean waiting time of any queueing system, and furthermore, this calculation depends only on mean values and not on distributions.

The first rigorous proof of this deceptively simple result did not appear until 1961 [LITT 61], and since then there have been a plethora of different proofs. Perhaps the simplest proof, albeit a somewhat heuristic one, is given below in the context of a discrete-time system.

*Proof of Little's result:* Begin by assuming the mean values $L$ and $W$ exist and that the queueing system is in equilibrium. Consider a long interval of $k$ slots $\{n = 0, 1, \ldots, k-1\}$, and let the mean arrival rate be $\lambda$ customers per slot. Then the mean number of customers that arrive is $\lambda k$. Let each customer bring its waiting time with it. Then the mean amount of waiting time brought in during the interval is $\lambda k W$.

Now each customer in the system uses up waiting time linearly with time. So if $L$ is the mean number of customers present during the interval, then $Lk$ is the mean number of slots used up during the $k$ slot interval. If we let $k \to \infty$ we can expect the amount of waiting time brought into the queueing system to be equal to the amount of waiting time used up. Thus

$$\lim_{k \to \infty} \frac{\lambda k W}{Lk} = 1 \tag{4.1}$$

and so

$$L = \lambda W \tag{4.2}$$

Note that this proof did not require any special assumptions concerning the arrival and service process or the service discipline.

## 4.2 DISCRETE-TIME QUEUEING CONVENTIONS

In all the queueing systems that follow, we shall assume that arrivals occur just after the beginning of a slot, while departures take place just before the end of a slot. The number of arrivals in slot $n$ will be denoted by $\{a_n, n = 0, 1, 2, \ldots\}$, where $\{a_n\}$ is a sequence of i.i.d. random variables with a specified distribution. The number of departures in slot $n$ will be denoted by $\{d_{n+1}, n = 0, 1, 2, \ldots\}$, with $d_0 \equiv 0$. The queue length process at the slot boundaries will be denoted by $\{y_n, \ n = 0, 1, 2, \ldots\}$, with $y_0$ arbitrary. In this case we have

$$y_{n+1} = y_n + a_n - d_{n+1} \tag{4.3}$$

The queue length process observed just after the arrivals have taken place will be denoted by $\{x_n, n = 0, 1, 2, \ldots\}$, where

$$x_n = y_n + a_n \tag{4.4}$$

Using equations (4.3) and (4.4), then

$$x_{n+1} = x_n + a_{n+1} - d_{n+1} \tag{4.5}$$

Figure 4.1 should help to make the above conventions clear.

## 4.3 DISCRETE-TIME *M/M/1* QUEUE

In this queue, arrivals form an independent Bernoulli process, with $a_n \in \{0, 1\}$, $n = 1, 2, 3, \ldots$, and there is unlimited waiting room. The queueing discipline is first-come first-served.

Let the probability of an arrival in a slot be $\alpha$ and the probability of a departure in a slot be $\beta$, and assume that the queueing system is in equilibrium. Then the state transition diagram is shown in Fig. 4.2, and the queue length process $\{x_n, n = 0, 1, 2, \ldots\}$ is a Markov chain with infinite state space. It is not difficult to show that this Markov chain is irreducible, aperiodic and recurrent non-null, and thus has a unique stationary probability distribution, provided that the mean rate of arrivals per slot, $\alpha$, is strictly less than the mean rate at which packets can be served per slot, $\beta$. This should be intuitively obvious in that if $\alpha > \beta$ the queue length

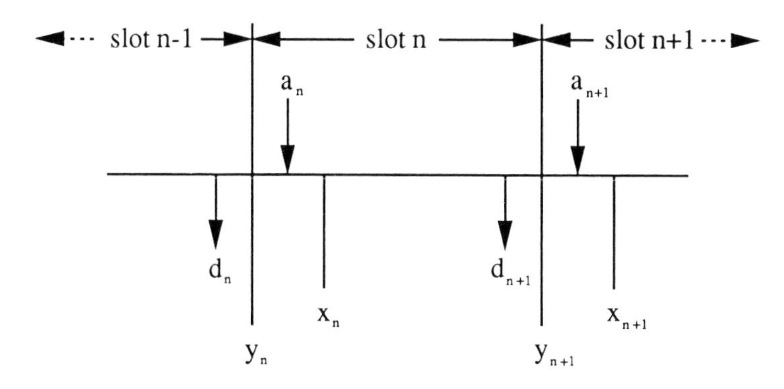


*Figure 4.1 Conventions for arrivals, departures and state of a discrete-time queue*

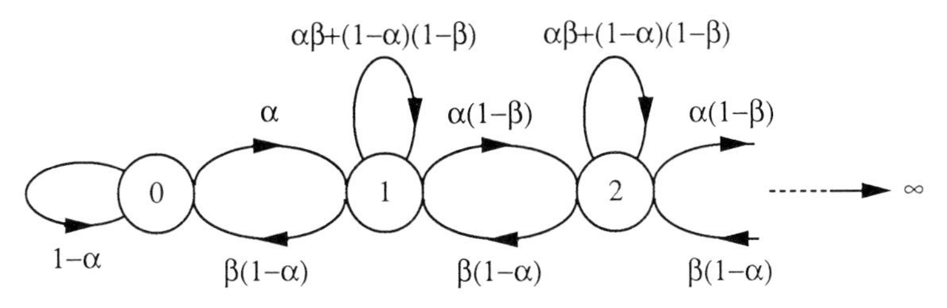


*Figure 4.2 State transition diagram for a discete-time M/M/1 queue*

process will become unstable and the number of customers in the system will build up to infinity. The case when $\alpha = \beta$ is somewhat pathological, and will not be considered further here.

Then under the assumption that $\alpha < \beta$, we can express the balance equations of the discrete-time $M/M/1$ queue as follows:

$$\Pi_0 = \Pi_0(1-\alpha) + \Pi_1\beta(1-\alpha) \tag{4.6}$$

$$\Pi_1 = \Pi_0\alpha + \Pi_1[\alpha\beta + (1-\alpha)(1-\beta)] + \Pi_2[(1-\alpha)\beta] \tag{4.7}$$

$$\Pi_2 = \Pi_1[\alpha(1-\beta)] + \Pi_2[\alpha\beta + (1-\alpha)(1-\beta)] + \Pi_3[(1-\alpha)\beta] \tag{4.8}$$

or, in general

$$\Pi_K = \Pi_{K-1}[\alpha(1-\beta)] + \Pi_K[\alpha\beta + (1-\alpha)(1-\beta)] + \Pi_{K+1}[(1-\alpha)\beta] \qquad K = 2, 3, 4, \ldots \tag{4.9}$$

Solving these equations recursively, we get

$$\Pi_1 = \Pi_0 \frac{\alpha}{\beta(1-\alpha)} \tag{4.10}$$

$$\Pi_2 = \Pi_0 \frac{\alpha^2(1-\beta)}{\beta^2(1-\alpha)^2} \tag{4.11}$$

$$\Pi_3 = \Pi_0 \frac{\alpha^3(1-\beta)^2}{\beta^3(1-\alpha)^3} \tag{4.12}$$

or, in general

$$\Pi_K = \Pi_0 \frac{\alpha^K(1-\beta)^{K-1}}{\beta^K(1-\alpha)^K} \qquad K = 1, 2, 3, \ldots \tag{4.13}$$

Writing

$$\gamma = \frac{\alpha(1-\beta)}{\beta(1-\alpha)} \tag{4.14}$$

then

$$\Pi_K = \Pi_0 \frac{\gamma^K}{1-\beta}, \qquad K = 1, 2, 3, \ldots \tag{4.15}$$

Using the normalising equation

$$\sum_{K=0}^{\infty} \Pi_K = 1 \tag{4.16}$$

then equations (4.15) and (4.16) give

$$\Pi_0\left[1 + \frac{1}{1-\beta}\sum_{K=1}^{\infty} \gamma^K\right] = 1 \tag{4.17}$$

This can be written as

$$\Pi_0 \frac{1}{1-\beta}\left[\sum_{K=0}^{\infty} \gamma^K - \beta\right] = 1 \tag{4.18}$$

We thus obtain $\Pi_0$ as

$$\Pi_0 = \frac{(1-\beta)(1-\gamma)}{1-\beta(1-\gamma)} \tag{4.19}$$

The final result for the equilibrium distribution is thus obtained from equations (4.15) and (4.19) as

$$\Pi_K = \frac{(1-\gamma)\gamma^K}{1-\beta(1-\gamma)}, \qquad K > 0 \tag{4.20}$$

with $\Pi_0$ given by equation (4.19).

The mean normalised throughput of this queue is given by the fraction of time the server is busy. That is

$$S = \sum_{K=1}^{\infty} \Pi_K = 1 - \Pi_0 \tag{4.21}$$

Substituting for $\Pi_0$ from equation (4.19) we get

$$S = \frac{\gamma}{1-\beta(1-\gamma)} \tag{4.22}$$

Substituting for $\gamma$ from equation (4.13) then

$$S = \frac{\alpha}{\beta} \tag{4.23}$$

This normalised quantity is in units of customers per mean customer service time. In units of customers per slot then

$$S = \alpha \qquad \textit{customers/slot} \tag{4.24}$$

That is, with $\alpha < \beta$ and the system in equilibrium, the mean rate per slot at which customers pass through the system is equal to the mean rate of customer arrivals per slot, as we should expect.

The mean waiting time can be found directly from Little's result as

$$W = \frac{\sum_{K=0}^{\infty} K\Pi_K}{\alpha} \qquad \textit{slots} \tag{4.25}$$

This gives

$$W = \left[\sum_{K=1}^{\infty} K\gamma^{K-1}\right]\frac{\gamma(1-\gamma)}{1-\beta(1-\gamma)}\frac{1}{\alpha}$$

$$= \frac{\gamma}{(1-\gamma)[1-\beta(1-\gamma)]}\frac{1}{\alpha}$$

or

$$W = \frac{1}{(1-\gamma)\beta} \qquad \textit{slots} \tag{4.26}$$

Normalising to mean customer service time, then

$$W = \frac{1}{1-\gamma} \tag{4.27}$$

It is interesting to note that if the slot length becomes very small so that $\alpha \to 0$ and $\beta \to 0$, in such a way that $\alpha/\beta \to \rho$, which is a constant, the equilibrium distribution $\Pi_K$ becomes

$$\Pi_K = (1-\rho)\rho^K, \qquad K = 0, 1, 2, \ldots \tag{4.28}$$

For those familiar with continuous-time queueing theory, equation (4.28) will be recognised as the equilibrium distribution of the continuous-time $M/M/1$ queue, where $\rho$ is the system utilisation, defined as the ratio of the mean arrival rate, $\lambda$, to the mean service rate, $\mu$.

Also interesting in the context of a comparison with the continuous-time $M/M/1$ queue is a comparison of waiting time. In discrete-time, equation (4.26) can be manipulated to give

$$W = \frac{1-\alpha}{\beta-\alpha} \tag{4.29}$$

A continuous-time $M/M/1$ queue with mean arrival rate $\alpha$ and mean service rate $\beta$ has waiting time given as [KLEI 75]

$$W = \frac{1}{\beta-\alpha} \tag{4.30}$$

Thus the mean waiting time in the discrete-time $M/M/1$ queue is always less than that of the equivalent continuous system; they differ by a factor of $1-\alpha$.

Equivalent results to those obtained in this section can be found in [BHAR 80].

## 4.4 DISCRETE-TIME $M/M/1/J$ QUEUE

This queue is similar to the discrete-time $M/M/1$ queue, but has a finite waiting room of $J$ customers, including any in service. The state transition diagram is given in Fig. 4.3. The queue length process $\{x_n; n = 0, 1, 2, \ldots\}$ is a Markov chain with a finite state space and clearly satisfies the conditions to have a unique stationary probability distribution. It should also be clear that a solution of the balance equations will differ only in the evaluation of $\Pi_0$ from the normalising equation, in that the latter is now the summation over a finite number of states. Thus, using equation (4.15) we have

$$\Pi_K = \Pi_0 \frac{\gamma^K}{1-\beta}, \qquad K = 1, 2, 3, \ldots, J \tag{4.31}$$

with $\gamma$ given by equation (4.14), as before. The normalising equation now becomes

$$\sum_{K=0}^{J} \Pi_K = 1 \tag{4.32}$$

Then using equations (4.31) and (4.32)

$$\Pi_0 \frac{1}{1-\beta}\left[\sum_{K=0}^{J} \gamma^K - \beta\right] = 1 \tag{4.33}$$

or

$$\Pi_0 \frac{1}{1-\beta}\left[\frac{1-\gamma^{J+1}}{1-\gamma} - \beta\right] = 1 \tag{4.34}$$

We thus obtain $\Pi_0$ as

$$\Pi_0 = \frac{(1-\beta)(1-\gamma)}{(1-\gamma^{J+1}) - \beta(1-\gamma)} \tag{4.35}$$

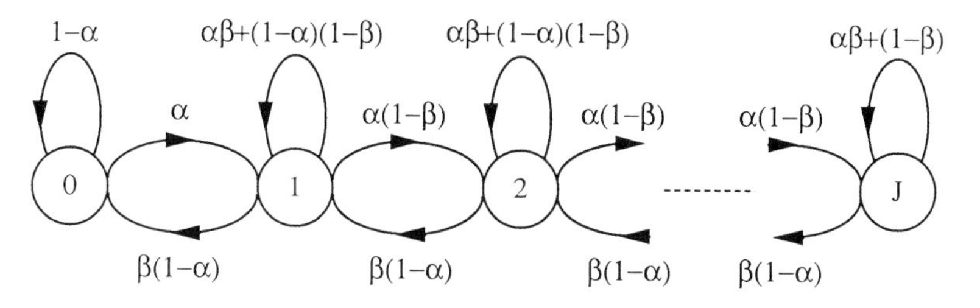


*Figure 4.3 State transition diagram for a discrete-time $M/M/1/J$ queue*

Finally, using equations (4.31) and (4.35)

$$\Pi_K = \frac{(1-\gamma)\gamma^K}{(1-\gamma^{J+1}) - \beta(1-\gamma)}, \qquad 1 \leq K \leq J \tag{4.36}$$

Throughput can be obtained in the same way as the discrete-time $M/M/1$ queue as

$$S = 1 - \Pi_0 \tag{4.37}$$

Substituting for $\Pi_0$ from equation (4.35) we get the normalised throughput as

$$S = \frac{\gamma(1-\gamma^J)}{(1-\gamma^{J+1}) - \beta(1-\gamma)} \tag{4.38}$$

As usual, this is in units of customers per mean customer service time. In units of customers per slot, we have

$$S = \frac{\gamma(1-\gamma^J)\beta}{(1-\gamma^{J+1}) - \beta(1-\gamma)} \qquad \textit{customers/slot} \tag{4.39}$$

Little's result can again be used to obtain the mean waiting time, but there is a subtle difference compared with the queue with unlimited waiting room. This is because there is now a finite probability that some of the arriving customers will be lost. Some thought will show that in this situation we must use the mean *throughput* in Little's result rather than the mean arrival rate, since this latter now includes customers that are turned away from the system. The mean waiting time, $W$, is thus given by

$$W = \frac{\sum_{K=0}^{J} K\Pi_K}{S} \tag{4.40}$$

Note that the evaluation of the numerator in equation (4.40), which is the mean queue size, involves the summation of a *finite* number of terms in an arithmetico-geometrical series. Practical communication networks obviously always have finite buffers, and it often turns out that the $\Pi_K$ have the form

$$\Pi_K = ax^K, \qquad 1 \leq K \leq J \tag{4.41}$$

where $a$ is a constant, and $|x| < 1$. In such cases (see Appendix)

$$\sum_{K=0}^{J} K\Pi_K = \frac{ax}{(1-x)^2}[1 - x^J[1 + (1-x)J]] \tag{4.42}$$

Using

$$x = \gamma \tag{4.43}$$

and

$$a = \frac{1-\gamma}{(1-\gamma^{J+1}) - \beta(1-\gamma)} \tag{4.44}$$

then equation (4.42) gives

$$\sum_{K=0}^{J} K\Pi_K = \frac{\gamma}{1-\gamma}\frac{[1-\gamma^J[1+J(1-\gamma)]]}{[(1-\gamma^{J+1}) - \beta(1-\gamma)]} \tag{4.45}$$

Substituting from equations (4.39) and (4.45) in equation (4.40), we obtain

$$W = \frac{[1-\gamma^J[1+J(1-\gamma)]]}{(1-\gamma)(1-\gamma^J)} \tag{4.46}$$

This is in units of mean customer service times. In units of slots, then

$$W = \frac{[1-\gamma^J[1+J(1-\gamma)]]}{(1-\gamma)(1-\gamma^J)\beta} \qquad \textit{slots} \tag{4.47}$$

In the above expressions for waiting time, it is assumed that $|\gamma| < 1$.

With a finite waiting room, we can also obtain the probability of blocking, $P_B$. This is just the fraction of time the system spends with the waiting room full, in which case any arriving customer will be turned away. Thus

$$P_B = \Pi_J \tag{4.48}$$

Using equation (4.36), then

$$P_B = \frac{(1-\gamma)\gamma^J}{(1-\gamma^{J+1}) - \beta(1-\gamma)} \tag{4.49}$$

## 4.5 DISCRETE-TIME $M^{a_n}/M/1$ QUEUE

We shall next generalise the discrete-time $M/M/1$ results to some extent by allowing for multiple arrivals in a slot. Let the arrivals at the beginning of each slot, $n$, be denoted by a sequence of i.i.d. random variables $\{a_n, n = 0, 1, 2, \ldots\}$ such that the probability of $K$ arrivals in slot $n$ is given by

$$\alpha_K = P(a_n = K), \qquad K = 0, 1, 2, \ldots \tag{4.50}$$

for all $n = 0, 1, 2, \ldots$. We retain the single geometric server, and assume unlimited waiting room. The condition for stability in this system is clearly

$$\sum_{K=0}^{\infty} K\alpha_K < \beta \tag{4.51}$$

Provided this condition holds, the Markov chain $\{x_n, n = 0, 1, 2, \ldots\}$ will have a unique stationary probability distribution.

In general, the transition probabilities can be specified as

$$p_{ij} = \begin{cases} 0 & : \quad j \leq i - 2 \\ \alpha_0 \beta & : \quad j = i - 1, i > 0 \\ \alpha_j & : \quad j \geq i = 0 \\ \alpha_{j-i+1}\beta + \alpha_{j-i}(1-\beta) & : \quad j \geq i, i > 0 \end{cases} \tag{4.52}$$

The balance equations for this Markov chain are

$$\Pi_0 = \Pi_0 \alpha_0 + \Pi_1[\alpha_0 \beta] \tag{4.53}$$

$$\Pi_1 = \Pi_1[\alpha_1 \beta + \alpha_0(1-\beta)] + \Pi_2[\alpha_0 \beta] + \Pi_0[\alpha_1] \tag{4.54}$$

$$\Pi_2 = \Pi_2[\alpha_1 \beta + \alpha_0(1-\beta)] + \Pi_3[\alpha_0 \beta] + \Pi_1[\alpha_2 \beta + \alpha_1(1-\beta)] + \Pi_0[\alpha_2] \tag{4.55}$$

$$\Pi_3 = \Pi_3[\alpha_1 \beta + \alpha_0(1-\beta)] + \Pi_4[\alpha_0 \beta] + \Pi_2[\alpha_2 \beta + \alpha_1(1-\beta)] + \Pi_1[\alpha_3 \beta + \alpha_2(1-\beta)] + \Pi_0(\alpha_3] \tag{4.56}$$

In general, we have

$$\Pi_K = \Pi_K[\alpha_1 \beta + \alpha_0(1-\beta)] + \Pi_{K+1}[\alpha_0 \beta] + \sum_{i=1}^{K-1} [\alpha_{K+1-i}\beta + \alpha_{K-i}(1-\beta)]\Pi_i + \Pi_0 \alpha_K \tag{4.57}$$

This reduces to

$$\Pi_K = \Pi_0 \alpha_K + \sum_{i=1}^{K+1} [\alpha_{K+1-i}\beta + \alpha_{K-i}(1-\beta)]\Pi_i \tag{4.58}$$

Equation (4.58) will hold for all $K = 0, 1, 2, \ldots$, provided we define

$$\alpha_j = 0, \qquad j < 0 \tag{4.59}$$

A rather elegant way of dealing with a set of equations like (4.58) is to transform them into a single equation by introducing the generating functions of the arrival and service processes, denoted by $A(z)$ and $B(z)$, respectively. $A(z)$ is the generating function of the number of customers arriving in a slot, which is given by

$$A(z) = \sum_{K=0}^{\infty} \alpha_K z^K \tag{4.60}$$

$B(z)$ is the generating function of the number of customers *being* served in a slot (with the queue non-empty). Thus a single customer is being served in a slot with probability $1 - \beta$, and service is completed in a slot with probability $\beta$. That is, the process is Bernoulli with probability $1 - \beta$

of success, and so $B(z)$ is given by

$$B(z) = \beta + (1 - \beta)z \tag{4.61}$$

The idea is to find the generating function of the queue length process, $\{x_n, n = 0, 1, 2, \ldots\}$, which is given by

$$P(z) = \sum_{K=0}^{\infty} \Pi_K z^K \tag{4.62}$$

Multiplying each of the equations (4.58) by $z^K$ and summing

$$P(z) = \sum_{K=0}^{\infty} \alpha_K z^K \Pi_0 + \sum_{K=0}^{\infty} z^K \left[ \sum_{i=1}^{K+1} [\alpha_{K+1-i}\beta + \alpha_{K-i}(1 - \beta)]\Pi_i \right]$$

$$= A(z)\Pi_0 + \sum_{i=1}^{\infty} \sum_{K=i-1}^{\infty} [(\alpha_{K+1-i}\beta + \alpha_{K-i}(1 - \beta))z^K \Pi_i] \tag{4.63}$$

Manipulating this, we find

$$P(z) = A(z)\Pi_0 + \left[ \frac{P(z) - \Pi_0}{z} \right] A(z)B(z) \tag{4.64}$$

This results in

$$P(z) = \Pi_0 A(z) \left[ \frac{z - B(z)}{z - A(z)B(z)} \right] \tag{4.65}$$

Since from equation (4.62)

$$P(1) = \sum_{K=0}^{\infty} \Pi_K = 1 \tag{4.66}$$

substituting $z = 1$ in equation (4.65), and noting from equation (4.61) that $B(1) = 1$, an indeterminacy of the form 0/0 results, which can be resolved using l'Hospital's rule to give:

$$\Pi_0 = 1 - A^{(1)}(1)/\beta \tag{4.67}$$

(Recall that $A^{(n)}(1)$ is the terminology used to denote the $n$th derivative of $A(z)$ evaluated at $z = 1$.) The desired generating function is thus given by

$$P(z) = \frac{A(z)(z - B(z))(1 - A^{(1)}(1)/\beta)}{z - A(z)B(z)} \tag{4.68}$$

Note from the properties of the generating function that $A^{(1)}(1)$ is the mean number of arrivals per slot. That is

$$A^{(1)}(1) = \sum_{K=0}^{\infty} K\alpha_K \tag{4.69}$$

This will therefore depend on the distribution of the random variables $\{a_n, n = 0, 1, 2, \ldots\}$.

The normalised throuput of the system is given by

$$S = 1 - \Pi_0 \tag{4.70}$$

which, from equation (4.67) is

$$S = A^{(1)}(1)/\beta \tag{4.71}$$

and, in terms of customers per slot is

$$S = A^{(1)}(1) \qquad \textit{customers/slot} \tag{4.72}$$

To find the mean waiting time via Little's result, we must first evaluate the mean queue length. This can be obtained from the generating function as

$$\sum_{K=0}^{\infty} K\Pi_K = P^{(1)}(1) \tag{4.73}$$

Using equation (4.68), then again an indeterminacy of the form 0/0 arises, and so l'Hospital's rule must be used to resolve this.

Define

$$P(z) = \frac{u(z)}{v(z)} \tag{4.74}$$

with $u(1) = v(1) = 0$. Hence

$$P^{(1)}(z) = \frac{v(z)u^{(1)}(z) - u(z)v^{(1)}(z)}{v(z)^2} \tag{4.75}$$

This is clearly of the form 0/0 at $z = 1$. Applying l'Hospital's rule once, we have

$$P^{(1)}(z) = \frac{v(z)u^{(2)}(z) - u(z)v^{(2)}(z)}{2v(z)v^{(2)}(z)} \tag{4.76}$$

This still evaluates to 0/0, and so a second application of the rule gives

$$P^{(1)}(z) = \frac{v^{(1)}(z)u^{(2)}(z) + v(z)u^{(3)}(z) - u^{(1)}(z)v^{(2)}(z) - u(z)v^{(3)}(z)}{2v(z)v^{(3)}(z) + 2v^{(1)}(z)^2} \tag{4.77}$$

Since $u(1) = v(1) = 0$, then

$$P^{(1)}(1) = \frac{v^{(1)}(1)u^{(2)}(1) - u^{(1)}(1)v^{(2)}(1)}{2v^{(1)}(1)^2} \tag{4.78}$$

Substituting the appropriate values, and since $B^{(2)}(1) = 0$, we get

$$P^{(1)}(1) = A^{(1)}(1) + \frac{A^{(2)}(1) + 2A^{(1)}(1)B^{(1)}(1)}{2(1 - A^{(1)}(1) - B^{(1)}(1))} \tag{4.79}$$

or

$$P^{(1)}(1) = A^{(1)}(1) + \frac{A^{(2)}(1) + 2A^{(1)}(1)(1-\beta)}{2(\beta - A^{(1)}(1))} \tag{4.80}$$

Depending on the distribution of the $\{a_n,\ n = 0, 1, 2, \ldots\}$, the mean queue length, and hence the mean waiting time can now be evaluated. Some examples follow.

*Example 4.1*

*Bernoulli arrivals:* If $a_n = a \in \{0, 1\}$, $n = 0, 1, 2, \ldots$, then the $\{a_n, n = 0, 1, 2, \ldots\}$ form an independent Bernoulli stream. This is therefore just a discrete-time $M/M/1$ queue, as considered in Section 4.4. In this case

$$A(z) = (1 - \alpha) + \alpha z \tag{4.81}$$

Hence $A^{(1)}(1) = \alpha$ and $A^{(2)}(1) = 0$. Substituting in equation (4.80) then

$$P^{(1)}(1) = \frac{\alpha(1-\alpha)}{\beta - \alpha} \tag{4.82}$$

Then, by Little's result,

$$W = \frac{1-\alpha}{\beta - \alpha} \tag{4.83}$$

This is exactly the result we obtained in equation (4.29), as we should expect.

*Example 4.2*

*Poisson arrivals:* If $\{a_n, n = 0, 1, 2, \ldots\}$ is a sequence of i.i.d. Poisson random variables, then we can use equation (2.136) to write $A(z)$ as

$$A(z) = e^{\lambda(z-1)} \tag{4.84}$$

where $\lambda$ is the mean number of arrivals per slot.

Then $A^{(1)}(1) = \lambda$ and $A^{(2)}(1) = \lambda^2$. Substituting these values in equation (4.80) gives

$$P^{(1)}(1) = \frac{\lambda(2-\lambda)}{2(\beta - \lambda)} \tag{4.85}$$

With the condition for stability, $\lambda < \beta$, satisfied, Little's result gives the waiting time as

$$W = \frac{2-\lambda}{2(\beta-\lambda)} = \frac{1-\lambda/2}{\beta - \gamma} \tag{4.86}$$

It is of interest to compare the mean waiting time of the discrete-time $M^{a_n}/M/1$ queue with $\{a_n\}$ a Bernoulli process, with that of the mean waiting time when $\{a_n\}$ is Poisson distributed, when the mean arrival rates are the same.

For the case of Bernoulli arrivals let

$$W = W_B = \frac{1-\alpha}{\beta-\alpha} \tag{4.87}$$

For the case of Poisson arrivals let

$$W = W_P = \frac{1-\alpha/2}{\beta-\alpha} \tag{4.88}$$

We find that

$$\frac{W_P}{W_B} = \left[\frac{1-\alpha/2}{1-\alpha}\right] \tag{4.89}$$

Clearly

$$W_P > W_B \tag{4.90}$$

with

$$\lim_{\alpha \to 0} W_P = W_B \tag{4.91}$$

Thus the mean waiting time of the Poisson system is always greater than that of the Bernoulli system, with the two tending to the same value as the mean arrival rate approaches zero. This is a consequence of the fact that the variance of the Poisson distribution, which in this case is $\lambda = \alpha$, is greater than that of the Bernoulli distribution, which is $\alpha(1-\alpha)$. It is a general result that determinism minimises delays, and a deterministic system, which has a variance of zero, gives the best delay characteristics.

Similar results to those obtained here can be found in [KING 90].

## 4.6 DISCRETE-TIME $M^{a_n}/M^{d_n}/\infty$ QUEUE

The next discrete-time queue we shall examine is the $M^{a_n}/M^{d_n}/\infty$ queue. This queue again has geometrically distributed interarrival times, and geometrically distributed service times, but has both batch arrivals and batch departures in a slot. It also has an infinite number of servers, which is something of a quantum leap from the previous single server systems. Although this system can be considered as a queue, clearly any arriving customer must always find a server available, and will thus pass through the queue in a time that is exactly equal to the geometrically distributed service time. The queue is therefore seen by an arriving customer as a geometrically distributed delay. This feature makes the queue relatively easy to analyse compared to a similar queue with a finite number of servers; in this latter case, a large number of state transitions would have to be explicitly taken into consideration in formulating the balance equations of the corresponding Markov chain.

So far we have considered two different approaches to solving the balance equations; namely, recursion, and the use of generating functions. Sometimes it is possible to bypass the whole process of writing down and solving the balance equations when finding an equilibrium distribution by using the concept of *time reversal.* This method is centred around the use of *Kelly's Lemma* [KELL 79], which is of sufficient importance to be stated again:

*Kellys's Lemma:*

If $\Pi$ is a distribution, and $P$ and $P'$ are two transition matrixes on $\mathbf{X}$ such that

$$\Pi_i p_{ij} = \Pi_j p'_{ji}, \qquad i, j \in \mathbf{X} \tag{4.92}$$

with

$$P = [p_{ij}, \qquad i, j \in \mathbf{X}]$$

$$P' = [p'_{ji}, \qquad i, j \in \mathbf{X}]$$

then $\Pi P = \Pi$ and $P'$ is the transition matrix of $(\Pi, P)$ *reversed in time*, $(\Pi, P)$ being the Markov chain with equilibrium distribution $\Pi$ and transition matrix $P$.

Note that this is not a deductive method of finding an equilibrium distribution in that one first requires to make some assertion concerning the form of the equilibrium distribution, and then show that this satisfies equation (4.92). In many cases the form of the equilibrium distribution can be *guessed* by heuristic reasoning.

We shall illustrate the method by finding the equilibrium distribution of the discrete-time $M^{a_n}/M^{d_n}/\infty$ queue for the case where $\{a_n, n = 0, 1, 2, \ldots\}$ is a sequence of i.i.d. Poisson random variables with mean $\lambda$.

With reference to the conventions in Section 4.2, then the queue length process $\{y_n, n = 0, 1, 2, \ldots\}$, observed at the slot boundaries is defined by

$$y_{n+1} = y_n + a_n - d_{n+1} \tag{4.93}$$

where $\{a_n, n = 0, 1, 2, \ldots\}$ is the Poisson arrival sequence with rate $\lambda > 0$, and $\{d_n\}$ is the departure sequence, with $x_0$ arbitrary and $d_0 \equiv 0$. The queue length process observed just after the arrivals have taken place is defined by

$$x_{n+1} = x_n + a_{n+1} - d_{n+1} \tag{4.94}$$

It will be assumed that each customer in the queue is served in a slot with probability $\beta$, independent of the other customers in the queue. Thus the probability of serving $j$ out of $i$ customers present in the queue just after the arrivals have taken place, $s(i, j)$, $0 \leq j \leq i$, is

$$s(i, j) = \binom{i}{j} \beta^j (1 - \beta)^{i-j} \tag{4.95}$$

Since this is a binomial distribution, we have

$$\sum_{j=0}^{i} s(i, j) = 1 \tag{4.96}$$

Then in any slot $n = 0, 1, 2, \ldots$, the number of departures, $d_{n+1}$, is obtained by an independent (Bernoulli) sampling of $x_n$.

To analyse the behaviour of the queue when in equilibrium, initially assume that $y_0 = 0$. Then in slot $n = 0$, $a_0$ Poisson arrivals join the $y_0 = 0$ already in the queue, and so $x_0$ is (trivially) Poisson. Just before the end of slot 0, $d_1$ customers are served and leave the queue. Thus, on average, $x_0\beta$ customers are served, and $x_0(1 - \beta)$ remain in the queue. This is a Bernoulli sampling of a Poisson process, and by the Poisson decomposition property $d_1$ is therefore Poisson. Similarly, $y_1$ must also be Poisson, and is joined in slot 1 by $a_1$ Poisson arrivals, so by the Poisson superposition property, $x_1$ is also Poisson. Clearly, by induction on $n$, $\{d_n, n = 1, 2, \ldots\}$ is a sequence of Poisson random variables which in equilibrium must have rate, $\lambda$, equal to that of the input process. Also, $\{x_n, n = 0, 1, 2, \ldots\}$ must also be Poisson, and in equilibrium, has a mean value, $a$, given by Little's result as

$$a = \lambda/\beta \tag{4.97}$$

That is, the mean number in the queue, $a$, is equal to the product of the arrival rate, $\lambda$, and the mean of the geometrically distributed service time, which is $1/\beta$.

Thus, the above results can be summarised by saying that, when the queue is in equilibrium, the departure sequence $\{d_n, n = 1, 2, \ldots\}$ is Poisson with rate $\lambda$, and the distribution of $\{x_n, n = 0, 1, 2, \ldots\}$ is given by

$$\Pi_K = \frac{a^K}{K!} e^{-a}, \qquad K = 0, 1, 2, \ldots \tag{4.98}$$

where $a = \lambda/\beta$.

Since $\{x_n, n = 0, 1, 2, \ldots\}$ is a Markov chain with a unique equilibrium distribution, then $\Pi_K$, given by equation (4.98), is independent of $y_0$. The above results therefore also hold for an arbitrary $y_0$.

It is interesting to note that $\{x_n\}$ is the sum of two Poisson processes $\{y_n\}$ and $\{a_n\}$, where, in equilibrium, $\{x_n\}$ has rate $\lambda/\beta$, and $\{a_n\}$ has rate $\lambda$. It follows that in equilibrium $\{y_n\}$ must have rate, $b$, given by

$$b = \lambda/\beta - \lambda = \frac{\lambda(1 - \beta)}{\beta} \tag{4.99}$$

Then the equilibrium distribution of $\{y_n, n = 0, 1, 2, \ldots\}$ is given by

$$\Pi_K = \frac{b^K}{K!} e^{-b}, \qquad K = 0, 1, 2, \ldots \tag{4.100}$$

with $b$ given by equation (4.99) above.

It is trivial that for this queue throughput, $S$, and mean waiting time, $W$, are given by

$$S = \lambda \tag{4.101}$$

$$W = 1/\beta \tag{4.102}$$

Although the reasoning that led to the above results was heuristic, the results can be formally verified by showing that they satisfy Kelly's lemma. For example, we can verify that

$$\Pi = \{\Pi_K, K = 0, 1, 2, \ldots\},$$

as given by equation (4.98) is the equilibrium distribution of $\{x_n, n = 0, 1, 2, \ldots\}$, by showing that this satisfies equation (4.92), as follows. The transition probabilities of the Markov chain $\{x_n\}$ are

$$P(x_{n+1} = j | x_n = i) = F(j + l - i) . s(i, l) \tag{4.103}$$

for all $i, j \in \mathbf{X}$, where

$$F(j + l - i) = \frac{\lambda^{j+l-i}}{(j + l - i)!} e^{-\lambda} = P(a_{n+1} = j + l - i) \tag{4.104}$$

and $s(i, l)$ is given by equation (4.95).

The transition probabilities of the Markov chain reversed in time are

$$P(x_n = i | x_{n+1} = j) = F(l) . s(j, j + l - i) \tag{4.105}$$

Substitution of the appropriate terms then verifies Kelly's lemma in that

$$\Pi_i F(j + l - i) . s(i, l) = \Pi_j F(l) . s(j, j + l - i) \tag{4.106}$$

Then $\Pi$ is the unique equilibrium distribution of $\{x_n\}$.

## 4.7 *S*-QUEUES

In this section a class of discrete-time queues called $S$-queues is defined and analysed. The treatment follows that of Walrand [WALR 83A]. It turns out that the discrete-time $M^{a_n}/M^{d_n}/\infty$ queue with $\{a_n\}$ Poisson (as in the previous section) is a special case of an $S$-queue. The motivation for studying $S$-queues is that they can be used to construct *networks of queues* whose equilibrium distribution has a particularly simple structure. Furthermore, these queueing networks can be interpreted as models for multiple access protocols of communication or computer networks.

Formally, an $S$-queue is defined as follows.

*Definition: S-queue*

Given an arbitrary sequence $\{a_n, n = 0, 1, 2, \ldots\}$ of $\{0, 1, 2, \ldots\}$-valued random variables, the queue length process of an $S$-queue is

$$y_{n+1} = y_n + a_n - d_{n+1}$$

where

$$P(d_{n+1} = j | y_m, d_m, 0 \leq m \leq n; a_k, k \geq 0, y_n + a_n = i) = s(i, j)$$

for $0 \leq j \leq i$ and $n \geq 0$, with $y_0$ arbitrary and $d_0 \equiv 0$. This definition considers the conventions of Section 4.2 apply.

We shall now consider that $\{a_n, n = 0, 1, 2, \ldots\}$ is a sequence of i.i.d. Poisson random variables with mean $\lambda$, and $s(i, j)$ takes the general form

$$s(i, j) = \begin{cases} c(i), & j = 0 \\ \dfrac{c(i)}{j!}\alpha(i)\alpha(i-1)\ldots\alpha(i-j+1), & j = 1, 2, \ldots, i \end{cases} \tag{4.107}$$

where

$$\alpha(k) = 1, \qquad k = 0$$

$$\alpha(k) > 0, \qquad k > 0$$

and $c(i)$ is such that

$$\sum_{j=0}^{i} s(i, j) = 1$$

Then provided a normalising constant, $c$, over the state space exists, the queue length process $\{y_n\}$ of an $S$-queue has an equilibrium distribution

$$\Pi_i = c\frac{\lambda^i}{\alpha(0)\ldots\alpha(i)}, \qquad i \geq 0 \tag{4.108}$$

Furthermore, in equilibrium the departure sequence $\{d_n, n = 1, 2, \ldots\}$ is Poisson with rate $\lambda$.

These results can be verified by considering the Markov chain $\{(y_n, d_n)\}$. From the definition of an $S$-queue, this has transition probabilities given by

$$P(y_{n+1} = j, d_{n+1} = l | y_n = i, d_n = k) = F(j + l - i) \,.\, s(j + l, l) \tag{4.109}$$

where $F(.)$ and $s(., .)$ are as previously defined. Time reversal arguments can then be used in a similar way to the previous section to verify that $\{(y_n, d_n)\}$ has an equilibrium distribution

$$P(y_n = y, d_n = d) = \Pi_y \,.\, F(d) \tag{4.110}$$

with $\Pi_y$ given by equation (4.108). In a single step this proves both that $\Pi$ is an equilibrium distribution and $\{d_n\}$ is Poisson.

To follow up the concept of time reversal pertaining to an $S$-queue, if we consider the arrivals, queue length process and departures at successive times as a Markov chain $\{(a_n, y_n, d_n)\}$, then this has the same stationary law as $\{(d_n, y_n, a_n)\}$ reversed in time. Also, when in equilibrium, the random variables $\{a_m, y_n, d_l, 1 \leq n \leq m\}$ are independent.

To prove these statements, let

$$v_n = a_n, y_n, d_n)$$

$$w_n = (d_n, y_n, a_n)$$

$$v = (a, y, d)$$

$$v' = (a', y', d')$$

We shall show that, in equilibrium

$$P(v_n = v, v_{n+1} = v') = P(w_n = v', w_{n+1} = v)$$

That is

$$P(v_n = v) . P(v_{n+1} = v' | v_n = v) = P(w_n = v') . P(w_{n+1} = v | w_n = v') \tag{4.111}$$

Showing equation (4.111) holds will imply the statements pertaining to reversibility and independence of $\{(d_n, y_n, a_n)\}$.

Equation (4.111) is equivalent to

$$\begin{aligned} &P(y_n = y, a_n = a) . P(y_{n+1} = y', d_{n+1} = d' | y_n = y, a_n = a) \\ &\quad = P(a_n = d', y_n = y') . P(y_{n+1} = y, d_{n+1} = a | a_n = d', y_n = y') \end{aligned} \tag{4.112}$$

Using equation (4.110), then equation (4.112) becomes

$$\begin{aligned} &\Pi_y F(a) s(y + a, d') 1\{y' + d = y + a\} \\ &\qquad = \Pi_{y'} F(d') s(y' + d', a) 1\{y + d = y' + d'\} \end{aligned} \tag{4.113}$$

where the notation $1\{. = .\}$ denotes a function that has value 1 if the statement contained within the braces is true, otherwise the value of the function is 0.

Using equations (4.107) and (4.108) in (4.113) establishes that equality and thus completes the verification.

The fact that the random variables $\{a_m, y_n, d_l, l \leq n \leq m\}$ are independent means that for an $S$-queue, at any time $n$, past departures, present state and future arrivals are independent. This input-output property of queues is known as *quasi-reversibility*. A more formal definition follows.

*Quasi-reversibility:* Consider an $S$-queue with Poisson arrival sequence, as previously defined. The queue is said to be quasi-reversible if it has an equilibrium distribution under which $\{d_n, n = 1, 2, \ldots\}$ is a Poisson sequence such that for all $n$, $\{d_l, l \leq n\}$ and $y_n$ are independent.

We shall now show that an $S$-queue is quasi-reversible if and only if $s(i, j)$ is as in equation (4.107).

Only the necessity has to be established, in which case the quasi-reversibility implies that the equilibrium distribution for $\{(y_n, d_n)\}$ is of the form given by equation (4.110). Then using equation (4.109) we have

$$l! \sum_{i=0}^{n} \Pi_i \frac{\lambda^{n-l-i}}{(n-i)!} s(n, l) = \Pi_{n-l} \tag{4.114}$$

Dividing term by term by the same equation for $l = 0$ then shows that

$$\frac{l!}{\lambda} \frac{s(n, l)}{s(n, 0)} = \frac{\Pi_{n-l}}{\Pi_n} \tag{4.115}$$

Defining

$$\alpha(0) = 1$$

$$\alpha(i+1) = \frac{\lambda \Pi_i}{\Pi_{i+1}}, \qquad i \geq 0$$

then we have

$$\Pi_i = c \frac{\lambda^i}{\alpha(0) \ldots \alpha(i)}$$

with

$$c = \Pi_0$$

With this definition, equation (4.115) becomes

$$\frac{l!}{\lambda} \frac{s(n, l)}{s(n, 0)} = \frac{\alpha(n) \ldots \alpha(n-l+1)}{\lambda} \tag{4.116}$$

Writing $n = i$, $l = j$ we have

$$s(i, j) = \frac{s(i, 0)}{j!} \alpha(i) \ldots \alpha(i-j+1)$$

which is equivalent to equation (4.107).

To give some examples of $S$-queues, consider the following.

*Example 4.3:*

*Geometric, infinite server system:* Let

$$\alpha(i) = \frac{\beta}{1-\beta} i, \qquad i \geq 1$$

$$c(i) = (1 - \beta)^i$$

Then we have

$$s(i, j) = \binom{i}{j} \beta^j (1 - \beta)^{i-j}$$

and

$$\Pi_i = \frac{b^i}{i!} e^{-b}$$

with

$$b = \frac{\lambda(1 - \beta)}{\beta}$$

Note that this queue is exactly that defined previously by equations (4.99) and (4.100). That is, it is the queue length process of a discrete-time $M^{a_n}/M^{d_n}/\infty$ queue with $\{a_n\}$ Poisson observed just before the arrivals. The discrete-time $M^{a_n}/M^{d_n}/\infty$ queue with $\{a_n\}$ Poisson, expressed in terms of the process $\{y_n\}$, is thus a special case of an $S$-queue.

*Example 4.4:*

*Processor sharing discipline:* If we let

$$\alpha(i) = \mu$$

then we find

$$s(i, j) = c(i) \frac{\mu^j}{j!}, \qquad j = 0, 1, \ldots, i$$

In this case

$$\Pi_i = (1 - \rho)\rho^i$$

with $\rho = \lambda/\mu$.

Note that this has the same form as equation (4.28), the equilibrium distribution of the discrete-time $M/M/1$ queue when $\alpha \to 0$, $\beta \to 0$, and $\alpha/\beta \to \rho$.

Unlike the previous example, however, this doesn't mean that the discrete-time $M/M/1$ queue is an $S$-queue under the above conditions, but only that it has the same equilibrium distribution as the $S$-queue under consideration when the conditions hold. The behaviour of the two queues leading to that distribution is quite different, and the $S$-queue considered corresponds to a form of PS discipline in the sense that $s(i, j)$ is a strictly decreasing function of $i > j$, for a fixed $j$. That is, the service unit is divided equally amongst all the customers in the queue in each slot.

The observations that can be made from these examples and the foregoing theory are that a FCFS single server $S$-queue must be such that $s(i, 0)$ is a constant for $i \geq 1$, and so must be $s(i, 1)$ for $i \geq 2$. Considering the form of equation (4.107) then shows that no FCFS single server $S$-queue can be quasi-reversible. Extending this argument implies that *no FCFS $S$-queue with finitely many servers can be quasi-reversible.*

## 4.8 EXERCISES

4.1 Extend the discrete-time $M/M/1$ analysis in Section 4.3 for the case of two geometric servers, each operating at rate $\beta$. Find the equilibrium distribution of the resulting queue, and hence obtain expressions for the throughput and mean writing time. The stability condition $\alpha < 2\beta$ should be assumed.

4.2 Generalise the result of Exercise 4.1 to the case of $c > 1$ geometric servers, each operating at rate $\beta$, such that $\alpha < c\beta$. Note that this is a discrete-time $M/M^{d_n}/c$ system, with $d_n \in \{0, 1, \ldots, c\}$.

4.3 Extend the result of Exercise 4.2 for the case of a finite waiting room of $J$ customers; that is, a discrete-time $M/M^{d_n}/c/J$ system, $J > c$.

4.4 For the discrete-time $M^{a_n}/M/1$ queue in Section 4.5, find an expression for the mean waiting time when $\{a_n\}$ is a sequence of i.i.d. geometrically distributed random variables, with the probability of $K$ arrivals in a slot, $\alpha_K$, given by,

$$\alpha_K = \gamma^K(1 - \gamma)$$

where $\gamma = p/(1 - p)$, $0 < p < 0.5$. The stability condition $\gamma/(1 - \gamma) < \beta$ should be assumed. This arrival distribution is called an *extended Bernoulli distribution* [PUJO 91].

4.5 For the discrete-time $M^{a_n}/M/1$ queue in Section 4.5, assume that $\beta = 1$; the queue then becomes a discrete-time $M^{a_n}/D/1$ type. Find an expression for the mean waiting time in this queue when $\{a_n\}$ is a sequence of i.i.d. random variables with a binomial distribution with parameters $(N, p)$.

This queue is also called a discrete-time $Geo(N)/D/1$ type; analysis of a similar queue can be found in [DESM 90].

# 5. DISCRETE-TIME QUEUEING NETWORKS

All the queues that have been considered so far have been studied in isolation. In practice this is rarely the case, and the queues usually form part of a model for a larger system. For example, each user in a computer communication network might be modelled as a queue of messages. The messages can be waiting for a particular communication channel to become free, and so gain access to the network and be transmitted. Each time a new message is generated by a user, that user's queue length is increased by 1. Each time a message is transmitted, the appropriate queue length is decreased by 1. The entire network can be modelled by an interconnection of such queues, generally called a *queueing network*.

The simplest possible queueing network consists of two queues in tandem (series). For the case of $S$-queues, this situation is analysed in Section 5.1. The main result here is that due to the quasi-reversibility of $S$-queues, the equilibrium probability that there are $y^1$ customers at queue 1 and $y^2$ customers at queue 2 is simply the *product* of the equilibrium probability that there are $y^1$ customers at queue 1 and the equilibrium probability that there are $y^2$ customers at queue 2, when the queues are considered in isolation. In other words, the queues act *as if they are independent*. For obvious reasons, this result is known as a *product form* distribution.

*General networks of S-queues* are considered in Section 5.2. Here it is shown that the product form result carries through, provided that the routing probabilities (the probabilities of customers being routed between specific queues in the network, and from or to the outside of the network) are *independent*.

Section 5.3 introduces the basic *discrete-time queueing network model* for a communication network, and shows how this can be interpreted as a network of $S$-queues. The concept of *interfering queues* is discussed, where the service rate in one queue depends on the state of the other queues. Such models cannot be solved exactly, and approximations must be used.

An approximation technique known as *equilibrium point analysis* is introduced in Section 5.4; this can be used to handle the interfering queue problem, and solve models with *state dependent routing probabilities.*

Finally, Section 5.5 gives a generalisation of equilibrium point analysis to the case of multiple customer classes. This implies that customers in a queueing network can be given different statistical properties (service time distributions and routing probabilities) according to their class.

## 5.1 TANDEM *S*-QUEUES

The simplest possible network of *S*-queues is to consider two of the queues in tandem (series), as shown in Fig. 5.1, where the queues are represented by nodes.

The queue length process of queue, $i$, $i = 1, 2$, is

$$y^i_{n+i} = y^i_n + a^i_n - d^i_{n+1} \tag{5.1}$$

That is, we have the same definition as for an isolated *S*-queue, but with a superscript denoting the queue (or node) number. The arrivals at node 1, $\{a^1_n, n = 0, 1, 2, \ldots\}$, form a Poisson stream with rate $\lambda$. These arrivals are queued at node 1 (if necessary), served, and then passed on to node 2. Here, they are again queued (if necessary), served, and then leave the network.

To analyse this system, first note that if the queue represented by node 2 has infinite waiting room, node 1 must clearly behave as an isolated *S*-queue, since it is unaffected by what happens at node 2. Thus with the system in equilibrium, the probability that there are $i$ customers at node 1 is given by equation (4.108). Since for node 1 we have

$$y^1_{n+1} = y^1_n + a^1_n - d^1_{n+1}$$

and for node 2 we have

$$\begin{aligned} y^2_{n+1} &= y^2_n + a^2_n - d^2_{n+1} \\ &= y^2_n + d^1_n - d^2_{n+1} \end{aligned}$$

then since $d^1_n$ is an independent Poisson process (by quasi-reversibility of node 1), then the flow on the link between node 1 and node 2 is Poisson, and node 2 must also act like an *S*-queue in isolation. This implies that when in equilibrium the system output is Poisson with rate $\lambda$.

We thus conclude that the joint equilibrium probability of $i$ customers at node 1 and $j$ customers at node 2, $\Pi(i, j)$ is

$$\Pi(i, j) = \Pi_1(i)\Pi_2(j) \tag{5.2}$$

where $\Pi_1(i)$ and $\Pi_2(j)$ are the *marginal probabilities* that there are $i$ customers at node 1 and $j$ customers at node 2, respectively. These take the form given by equation (4.108), with a slight change in notation.

This is the *product form result* for discrete-time *S*-queues in tandem, and can clearly be extended to any number of queues by the quasi-reversibility of the nodes. That is, for $H$ queues in tandem, the

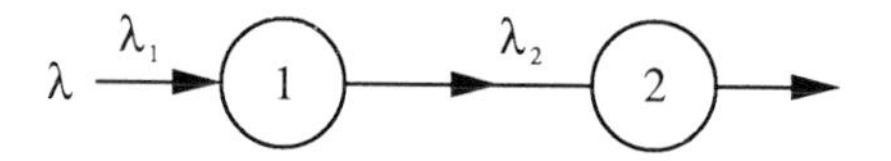


*Figure 5.1 Two discrete-time S-queues in tandem*

equilibrium probability of finding $i_1$ customers at node 1, $i_2$ customers at node 2,..., $i_H$ customers at node $H$, is

$$\Pi(i_1, i_2, \ldots, i_H) = \Pi_1(i_1)\Pi(i_2)\ldots\Pi_H(i_H) \tag{5.3}$$

Of course, more complex connections between the nodes can be allowed, so long as we maintain the Poisson nature of the flows. This is a direct consequence of the definition of quasi-reversibility and of the splitting and merging properties of the Poisson process. In other words, a network of $S$-queues in which each queue can have several input streams, and in which the output of any queue can be split, can still be analysed as if each queue were independent. The only restriction is that feedback must be impossible. That is, it must not be possible for customers which have left a queue ever to visit the same queue again in their passage through the network (feedback destroys the Poisson nature of the flows) [WALR 83B]. Such networks are called *feed-forward networks*, and tandem queues are special cases of these.

This logically leads us to ask what happens to the equilibrium distribution in a general network of discrete-time queues where feedback is allowed (and the flows on the links are no longer Poisson). For the case of $S$-queues, this general problem is addressed in the next section.

## 5.2 NETWORKS OF $S$-QUEUES

We next consider a general network of $S$-queues where any interconnection topology is allowed. To be more specific, for $i = 1, 2, \ldots, H$, let $\{y_n^i\}$ be the queue length process of an $S_i$-queue which has a Poisson arrival sequence of rate $\lambda_i$. Let $s(i, j)$ for this queue be as previously defined in (4.107), and let $\Pi_i(.)$, which is thus given by equation (4.108), be properly normalised.

These $H$ queues will be used to build a network as follows, where the previous conventions hold in that customers leaving a given node just before the end of the $n$th time slot reach their destination just after the start of the $n+1$th slot. Just before the end of slot $n - 1 \geq 0$, $d_n^i$ customers are served at node $i$, and are routed independently to node $j$, $i, j = 1, 2, \ldots, H$, with probability $r_{ij}$, or leave the network with probability

$$r_{i0} = 1 - \sum_{j=1}^{H} r_{ij}, \qquad i = 1, 2, \ldots, H \tag{5.4}$$

At time $n$, the customers routed to join node $j$ join the $y_n^j$ customers already there. At this same time $a_n^j$ customers also arrive at node $j$ from outside the network. The external arrival sequence is thus

$$\{a_n = (a_n^1, a_n^2, \ldots, a_n^H), \qquad n = 0, 1, 2, \ldots\} \tag{5.5}$$

It will be assumed that the

$$\{a_n^i, \quad i = 1, 2, \ldots, H\} \tag{5.6}$$

are *independent Poisson sequences* with respective rates $\gamma_i$. Then the aggregate external arrivals as given by (5.5) are Poisson with rate

$$\gamma = (\gamma_1, \gamma_2, \ldots, \gamma_H) \tag{5.7}$$

such that

$$\lambda_i = \gamma_i + \sum_{j=1}^{H} \lambda_j r_{ji} \qquad i = 1, 2, \ldots, H \tag{5.8}$$

These are the network's *flow conservation equations.*

With independent routing, as assumed here, these are *linear equations*, and a solution

$$\lambda = (\lambda_1, \lambda_2, \ldots, \lambda_H) \tag{5.9}$$

is a collection of possible *average rates of flow* of customers through the queues. In particular, if $\lambda_j$ is the average rate of customers through queue $j$, then

$$\sum_{j=1}^{H} \lambda_j r_{ji}$$

is the average rate of customers joining queue $i$ per slot from the other queues (including queue $i$). Thus the right hand side of equation (5.8) is the total average rate of customers going through queue $i$ per slot, i.e. $\lambda_i$.

An $S$-queue node in this general network is shown in Fig. 5.2. Now let

$$e_n = (e_n^1, e_n^2, \ldots, e_n^H) \tag{5.10}$$

where $e_n^i$ is the total number of customers entering node $i$ at time $n$. That is, $e_n^i$ consists of external arrivals $a_n^i$ and the customers coming from other nodes after being routed to node $i$. To be consistent with the definition of an (isolated) $S$-queue we have

$$y_{n+1}^i = y_n^i + e_n^i - d_{n+1}^i \tag{5.11}$$

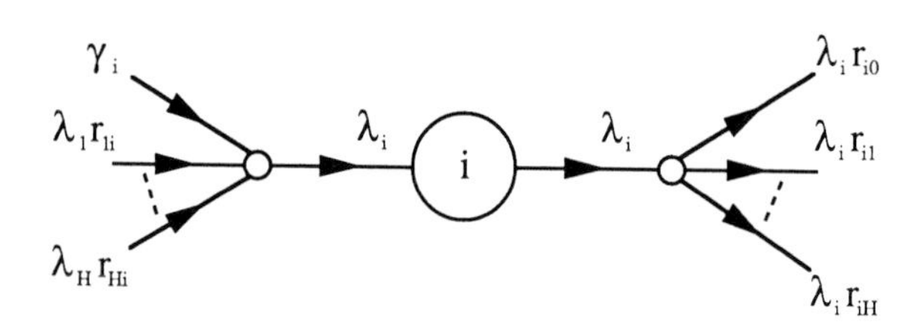


*Figure 5.2 S-queue node in a general queueing network*

and

$$P(d_{n+1}^i = l | y_m, a_m, e_m, 0 \le m \le n; y_n^i + e_n^i = k) = s_i(k, l) \tag{5.12}$$

where

$$y_m = (y_m^1, y_m^2, \ldots, y_m^H) \tag{5.13}$$

Also, let

$$b_n = (b_n^1, b_n^2, \ldots, b_n^H) \tag{5.14}$$

where $b_n^i$ is the number of customers leaving the network from node $i$ at time $n$.

Then we have the following key result:

*Product form theorem (Open networks)*

(a) An equilibrium distribution of

$$y_n = (y_n^1, y_n^2, \ldots, y_n^H)$$

is given by

$$\Pi(y^1, y^2, \ldots, y^H) = \Pi_1(y^1)\Pi_2(y^2)\ldots\Pi_H(y^H) \tag{5.15}$$

(b) In equilibrium, the departure process from the network is a Poisson sequence with rate

$$(\lambda_1 r_{10}, \lambda_2 r_{20}, \ldots, \lambda_H r_{H0})$$

(c) In equilibrium, the random variables

$$\{a_m, y_n, b_l, l \le n \le m\}$$

are independent for each $n$.

To prove the above, it is sufficient to show that for each $n \ge 0$, the random variables

$$\{y_n, e_n, a_m, b_l, l \le n < m\}$$

are independent in equilibrium and such that $\{e_n\}$ is a Poisson sequence with rate

$$\lambda = (\lambda_1, \lambda_2, \ldots, \lambda_H)$$

and that

$$\{y_n, a_m, b_l\}$$

is as stated in the theorem. This will prove statements (b) and (c), and the product form result (a) then follows directly by the properties of quasi-reversibility. To proceed by induction in time, assume the above to be true for some $n$, and let

$$f_n^i = e_n^i - a_n^i$$

and note that

$$f_n = (f_n^1, f_n^2, \ldots, f_n^H)$$

is obtained from $d_n$ by independent samplings with probabilities $r_{ij}$.

Then using the properties of an isolated $S$-queue that in equilibrium the departure sequence is Poisson, and that the random variables $\{a_m, y_n, d_l, l \leq n \leq m\}$ are independent we find that for the network as defined

$$\{y_{n+1}, d_{n+1}, a_m, b_l, l \leq n < m\}$$

are independent and all but $y_{n+1}$ are Poisson.

By the independence of the routing it follows that

$$\{y_{n+1}, f_{n+1}, a_m, b_l, l \leq n+1 \leq m\}$$

have the same property. Then the same must also be true for

$$\{y_{n+1}, e_{n+1}, a_m, b_l, l \leq n+1 < m\}$$

which establishes the proof.

A similar result can be established for *closed networks* of $S$-queues. For such networks, the external arrivals and departures are non existent (zero), and a fixed population of $N$ customers circulates among the queue nodes. The equivalent product form result for closed networks is as follows:

*Product form theorem (Closed networks)*

An equilibrium distribution of

$$y_n = (y_n^1, y_n^2, \ldots, y_n^H)$$

is given by

$$\Pi(y^1, y^2, \ldots, y^H) = G_N \sum_{i=1}^{H} r_i(y^i) \tag{5.16}$$

where

$$r_i(y^i) = \frac{\lambda_i^{y^i}}{\alpha(0)\alpha(1)\ldots\alpha_i(y^i)} \tag{5.17}$$

and

$$G_N = \frac{1}{\sum_{\mathbf{X}_N} \prod_{i=1}^{H} r_i(y^i)} \tag{5.18}$$

In the above, $G_N$ is a *normalising constant* to ensure that the equilibrium probabilities sum to one over the state space $\mathbf{X}_N$. Note also that with

$$y = (y^1, y^2, \ldots, y^H)$$

the state space for a closed network with $N$ customers is constrained such that

$$\sum_{i=1}^{H} y^i \leq N \tag{5.19}$$

That is

$$\mathbf{X}_N = \left\{ y \in \mathbf{X} \,\middle|\, \sum_{i=1}^{H} y^i \leq N \right\} \tag{5.20}$$

where

$$\mathbf{X} = \{0, 1, 2, \ldots\}^H$$

The inequality in (5.19) arises because of the way the queue length process $\{y_n\}$ is defined. That is, $\{y_n\}$ is the vector of queue lengths observed just after the departures and just before the arrivals, which means that $\{y_n\}$ is the *state of the customers who do not jump at time n*. Clearly then, $y$ can possibly take any combination of values within the constraint (5.19).

The product form result for closed networks can be easily proved in a similar way to the open networks, by induction in time, as follows.

The closed network is defined as the open one except that $a_n = b_n = 0$, $n \geq 0$. This also implies $f_n = e_n$, $n \geq 0$. Then assume that at some time $n$ the network is in equilibrium, and what leaves the nodes at this time, $d_n$, is a Poisson random variable with rate $\lambda = (\lambda_1, \lambda_2, \ldots, \lambda_H)$ such that $d_n$ and $y_n$ are independent. Then the claim is that for each $n \geq 0$, the random variables

$$\{y_n, e_n\}$$

are independent in equilibrium and such that $\{e_n\}$ is a Poisson sequence with rate

$$\lambda = (\lambda_1, \lambda_2, \ldots, \lambda_H)$$

Clearly, by the independence of the routing the claim is true for $n$. Then using the quasi reversibility properties of an (isolated) $S$-queue

$$\{y_{n+1}, d_{n+1}\}$$

are independent, and $d_{n+1}$ is Poisson with rate $\lambda$.

Again, by the independence of the routing

$$\{y_{n+1}, e_{n+1}\}$$

have the same property, which establishes the proof by induction. Then, the product form, suitably normalised on the restricted state space, will thus be invariant.

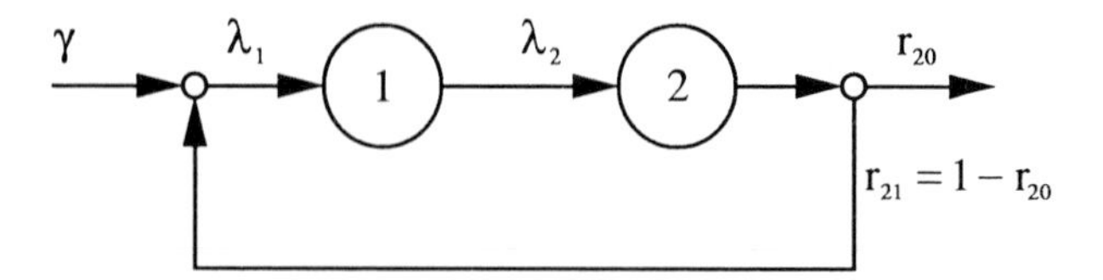


*Figure 5.3 Two S-queues in tandem with feedback*

*Example 5.1: Product form for S-queues*

Consider the system shown in Fig. 5.3, where the nodes are geometric, infinite server $S$-queues (defined in Section 4.8) with

$$\Pi_i(y^i) = \frac{b_i^{y^i}}{y^i!} e^{-b_i} \tag{5.21}$$

and

$$b_i = \frac{\lambda_i(1-\mu_i)}{\mu_i}, \qquad i = 1, 2 \tag{5.22}$$

External arrivals are at node 1 only, and are Poisson with rate $\gamma_1 = \gamma$. Customers are served at node 1 with rate $\mu_1$ (per server) and routed to node 2. Here they are served with rate $\mu_2$ (per server) and either leave the network with probability $r_{20}$, or are fed back to node 1 with probability $r_{21} = 1 - r_{20}$.

When $r_{21} = 0$, then the system becomes two $S$-queues in tandem without feedback. Equations (5.2) (or (5.15)), (5.21) and (5.22) then give the equilibrium distribution as

$$\Pi(y^1, y^2) = \Pi_1(y^1)\Pi_2(y^2) = \frac{b_1^{y^1} b_2^{y^2}}{y^1! y^2!} e^{-(b_1+b_2)}$$

$$= \frac{\left[\dfrac{\gamma(1-\mu_1)}{\mu_1}\right]^{y^1} \left[\dfrac{\gamma(1-\mu_2)}{\mu_2}\right]^{y^2}}{y^1! y^2!} e^{-\gamma\left[\frac{1}{\mu_1} + \frac{1}{\mu_2} - 2\right]} \tag{5.23}$$

If $0 < r_{21} < 1$ and feedback is allowed in the network of Fig. 5.3, then the equilibrium distribution is exactly the same as in equation (5.23), even though it is well known that feedback destroys the Poisson nature of flows on the links [WALR 83B].

The product form result for general networks of $S$-queues is therefore quite remarkable in that the nodes still behave independently, *as if* the flows on the links are Poisson. This is a result that is analogous to the well known Jackson's theorem for continuous time networks with Poisson arrivals, exponential service times, and independent routings [JACK 57].

When $r_{21} = 1$, then the system becomes a closed network. Equations (5.16), (5.17), (5.18) and (5.22) then give the equilibrium distribution as

$$\Pi(y^1, y^2) = \frac{\left[\frac{\lambda(1-\mu_1)}{\mu_1}\right]^{y^1}\left[\frac{\lambda(1-\mu_2)}{\mu_2}\right]^{y^2}}{y^1!y^2!} G_N \tag{5.24}$$

where

$$G_N = 1 \Bigg/ \sum_{y\in\mathbf{X}_N} \frac{\left[\frac{\lambda(1-\mu_1)}{\mu_1}\right]^{y^1}\left[\frac{\lambda(1-\mu_2)}{\mu_2}\right]^{y^2}}{y^1!y^2!} \tag{5.25}$$

In the above, $\lambda = \lambda_1 = \lambda_2$ is a solution to the flow conservation equations, which for a closed network become

$$\lambda_i = \sum_{j=1}^{H} \lambda_j r_{ji}, \qquad i = 1, 2, \ldots, H \tag{5.26}$$

In general, these are homogeneous equations and have infinitely many solutions, unlike equations (5.8) for the open networks where the external arrival rates $\gamma_i$, $i = 1, 2, \ldots, H$, give a fixed reference and hence a unique solution.

A solution to (5.26) is obtained by fixing one of the $\lambda_i$s arbitrarily and using this to determine the others. By fixing the same $\lambda_i$ to a different value will still yield the correct equilibrium distribution. In fact the only thing that changes is the normalising constant $G_N$. If this is recomputed appropriately, then the equilibrium probabilities will not change.

For example, say there are two customers circulating in the network and $\mu_1 = \mu_2 = 1/2$. We can fix $\lambda$ as $\lambda = \lambda_1 = \lambda_2 = 1$. Then we have from equation (5.24)

$$\Pi(y^1, y^2) = \frac{1}{y^1!y^2!} G_2$$

and

$$G_2 = \frac{1}{\sum_{y\in\mathbf{X}_2} \frac{1}{y^1!y^2!}}$$

where

$$\mathbf{X}_2 = \{y \in \mathbf{X} | y^1 + y^2 \leq 2\}$$

$$\mathbf{X} = \{0, 1, 2, \ldots\}^2$$

The normalising constant $G_2$ evaluates to 1/5, as shown in Table 5.1, which also shows the corresponding equilibrium probabilities.

**Table 5.1** EQUILIBRIUM DISTRIBUTION FOR THE CLOSED NETWORK CASE OF EXAMPLE 5.1

| $y^1$ | $y^2$ | $1/y^1!y^2!$ | $\Pi(y^1, y^2)$ |
|---|---|---|---|
| 0 | 0 | 1 | 1/5 |
| 0 | 1 | 1 | 1/5 |
| 1 | 0 | 1 | 1/5 |
| 1 | 1 | 1 | 1/5 |
| 0 | 2 | 1/2 | 1/10 |
| 2 | 0 | 1/2 | 1/10 |

Interestingly, if $\mu_1 = \mu_2 = 1$, then we find

$$\Pi(y^1, y^2) = \begin{cases} 1, & y^1 = y^2 = 0 \\ 0, & \text{otherwise} \end{cases}$$

This result is independent of the value of $N$, and means that irrespective of the network's population, the queues are always empty!

This result is easily explained in that what we are computing is the probability distribution of the customers who do not jump. With $\mu_1 = \mu_2 = 1$ then all customers must jump at the end of a slot, thus when the queue lengths are observed (the observation instant is just after the departures and just before the arrivals) all the customers are in flight between the queues.

When the state space is large, the evaluation of the normalising constant can be a major problem. A number of efficient algorithms for this evaluation are available, most notably the *convolution algorithm* and its variants [BUZE 73], [REIS 75], [REIS 81], [CONN 89]. Details of these can be found in almost any book on continuous-time queueing networks, and they will not be considered further here. A good exposition of the convolution algorithm can be found in [KING 90].

The product form results presented here are based on the ideas of Walrand [WALR 83A], [WALR 83B]. A number of other discrete-time product forms have been reported in the literature. These include [BHAR 80], [DADU 83], [PUJO 85], [DIJK 90], [PUJO 91], [HEND 90A] and [HEND 90B].

## 5.3 DISCRETE-TIME QUEUEING NETWORK MODELS FOR MULTIPLE ACCESS PROTOCOLS

Many computer communication networks use a single high bandwidth transmission channel through which all communication takes place. Examples are satellite channels in a wide area environment, and coaxial cable buses or optical fibre rings in a local environment, Irrespective of the type of topology of the transmission medium, such networks can be

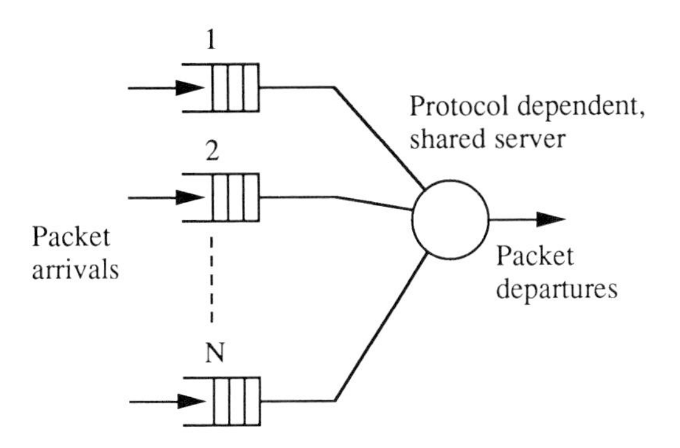


*Figure 5.4 Parallel FCFS queues with shared server (multiqueue)*

modelled as $N$ parallel FCFS queues serviced by a common pool of one or more shared servers, as shown in Fig. 5.4. Each of the queues represents a user $i$, $i = 1, 2, \ldots, N$. Each time user $i$ generates a message, this is placed in queue $i$ to await a server becoming free. The shared servers correspond to the channel and have a service discipline that is protocol dependent, so that, for example, in a polling network the (single) server is allocated to each queue in turn in a cyclic order. The state of a user is defined by the vector $(i_1, i_2, \ldots, i_k)$, where $i_m$, $1 \leq m \leq k$, is a non-negative integer that represents some particular feature of the system, such as the number of messages in the user's buffer, the remaining service time of the message currently being transmitted, and so on. This is known as a *multiqueue model.*

If we assume users to be statistically identical (a restriction to be relaxed later), an equivalent representation of the multiqueue model is as a closed, discrete-time queueing network. This alternative model consists of a $k$-dimensional array of $H$ queue nodes, where $H$ is finite. Each node $j = 1, 2, \ldots, H$ corresponds to a particular value of the vector $(i_1, i_2, \ldots, i_k)$. The nodes are populated by *users* (not messages) so that, for example, if $k = 2$, $i_1$ represents the number of messages in a user's buffer, and $i_2$ the remaining transmission time of the message at the head of the users queue, then all users at a given node $j$ that corresponds to the vector $(i_1, i_2) = (l, m)$ have $l$ messages in their buffer, and a remaining transmission time of $m$ slots for the message currently being transmitted.

If $x_n^j$ denotes the number of users at node $j$ during slot $n$, $n = 0, 1, 2, \ldots,$ then

$$x_n = (x_n^1, x_n^2, \ldots, x_n^H) \tag{5.27}$$

is a state vector of the network, and is a discrete-time Markov chain

$$\mathbf{x} = \{x_n = x, \qquad n = 0, 1, 2, \ldots, x \in \mathbf{X}_N\} \tag{5.28}$$

where

$$x = (x^1, x^2, \ldots, x^H) \tag{5.29}$$

and

$$\mathbf{X}_N = \left\{ x \in \mathbf{X} \,\middle|\, \sum_{j=1}^{H} x^j = N \right\} \tag{5.30}$$

with

$$\mathbf{X} = \{0, 1, 2, \ldots\}^H \tag{5.31}$$

Each node $j$ has batch arrivals and batch departures with arrival sequence $\{a_n^j, n = 0, 1, 2, \ldots\}$ of $\{0, 1, 2, \ldots\}$ valued random variables such that

$$x_{n+1}^j = x_n^j - d_{n+1}^j + a_{n+1}^j \tag{5.32}$$

where $\{d_n^j, n = 0, 1, 2, \ldots\}$ is the sequence of departures, from node $j$, with $d_0^j \equiv 0$. This definition considers that departures take place from a node just before the end of a slot, arrivals occur just after the beginning of a slot, and $\{x_n\}$ is the queue length process at all the nodes, observed just after the arrivals in slot $n$. Each user at node $i$ is served independently with probability $\mu_i$, and then routed to node $j$ with probability

$$r_{ij}, 0 \leq r_{ij}, \sum_{j=1}^{H} r_{ij} = 1 \qquad i, j = 1, 2, \ldots, H \tag{5.33}$$

It therefore follows that the probability of serving $r$ out of $t$ users present at node $i$ just after the arrivals, $s_i(t, r)$, is

$$s_i(t, r) = \binom{t}{r} \mu_i^r (1 - \mu_i)^{t-r} \tag{5.34}$$

Then each node can be interpreted as a queue of users with infinitely many servers, and a geometric service time distribution. In other words, when considered in isolation with a Poisson arrival sequence, each node is precisely a discrete-time $M^{a_n}/M^{d_n}/\infty$ queue of the type that was analysed in Section 4.6.

The nodes of these networks are thus special cases of $S$-queues, with the queue lengths observed just after rather than just before the arrivals. Allowing for this, the equilibrium distribution of the vector of queue lengths will thus have a product form provided that the routing probabilities are independent, as was shown in the preceding section. That is, an equilibrium distribution of

$$x_n = (x_n^1, x_n^2, \ldots, x_n^H)$$

is given by

$$\Pi(x^1, x^2, \ldots, x^H) = G_N \sum_{i=1}^{H} r_i(x^i) \tag{5.35}$$

where

$$r_i(x^i) = \frac{a_i^{x^i}}{x^i!} \tag{5.36}$$

with

$$a_i = \frac{\lambda_i}{\mu_i} \tag{5.37}$$

and

$$G_N = 1 \Bigg/ \sum_{\mathbf{X}_N} \prod_{i=1}^{H} r_i(x^i) \tag{5.38}$$

Here, $\lambda_i$ is the mean number of arrivals per unit time at node $i$, and is given by a suitable solution to the flow conservation equations

$$\lambda_i = \sum_{j=1}^{H} \lambda_j r_{ji} \qquad i = 1, 2, \ldots, H \tag{5.39}$$

Note that in the above, equation (5.36) relates to the queue lengths *after* the arrivals, and is obtained using equation (4.98).

To make the above ideas more concrete, consider the following example.

*Example 5.2: Product form discrete-time queueing network model for a multiple access protocol (2-user TDMA)*

Two users share a common transmission channel and each has buffer space for just one message. Messages are generated by each user with probability $\sigma$ per slot, with the message generation assumed to take place at the start of a slot. Slots are sufficiently large to accommodate exactly one fixed length message. The multiple access protocol assigns alternate slots to each of the users on a fixed, cyclic basis. The task is to find the mean throughput of the system using the product form result.

A discrete-time queueing network model for this system is shown in Fig. 5.5. This is really a special case of the familiar time division multiple access (TDMA) protocol [KUO 81], [STAL 85]. Each node in Fig. 5.5 represents (in isolation) a discrete-time $M^{a_n}/M^{d_n}/\infty$ queue. At any given time one of the users must be at node 2, and the other user at either node 1 or node 3. A user at node 1 or node 2 has no messages in its buffer, while a user at node 3 has one message in its buffer. A user at node 2 either generates a message with probability $\sigma$ and jumps to node 3 at the end of the slot, or does not generate a message (probability $1 - \sigma$), in which case its buffer remains empty, and jumps to node 1 at the end of the slot. A user at node 3 has one message in its buffer and can transmit this in the current slot and jump to node 2 at the end of the slot. A user at node 1 has no messages in its buffer, but can generate one at the start

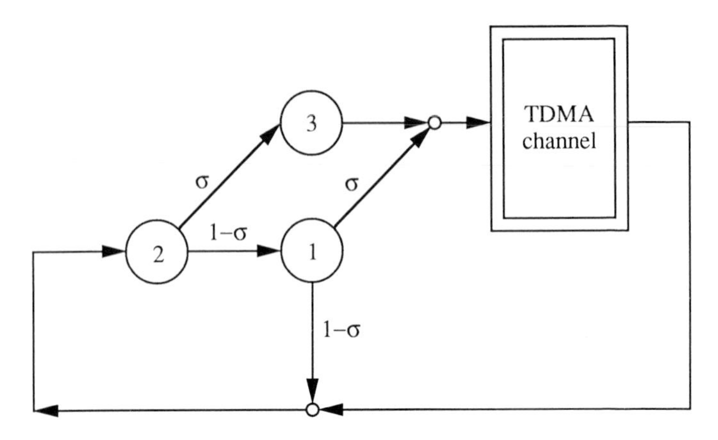


*Figure 5.5 Discrete-time queueing network model for two-user TDMA*

of the current slot with probability $\sigma$. In this case the message is immediately transmitted and the user jumps to node 2 at the end of the slot. If no message is generated at node 1, then the user jumps to node 2 at the end of the slot without transmitting.

The state vector of this network is

$$x = (x^1, x^2)$$

(since $x^3 = 1 - x^1$). The state space is

$$\mathbf{X}_2 = \{x \in \mathbf{X} | x^2 = 1, x^1 + x^3 = 1\}$$

$$\mathbf{X} = \{0, 1, 2, \ldots\}^2$$

The service rates at the nodes are

$$\mu_1 = \mu_2 = \mu_3 = 1$$

The non-zero routing probabilities are

$$r_{12} = r_{32} = 1, r_{23} = \sigma, r_{21} = 1 - \sigma$$

The flow conservation equations (5.39) become

$$\lambda_i = \sum_{j=1}^{3} \lambda_j r_{ji}, \qquad i = 1, 2, 3$$

These give

$$\lambda_1 = \lambda_2(1 - \sigma)$$

$$\lambda_2 = \lambda_1 + \lambda_3$$

$$\lambda_3 = \lambda_2 \sigma$$

Setting $\lambda_2 = 1$ then

$$\lambda_1 = 1 - \sigma$$

$$\lambda_3 = \sigma$$

Substitution of the $\lambda_i$, $\mu_i$, $i = 1, 2, 3$, in equation (5.38) yields

$$G_2 = 1$$

The equilibrium distribution is thus

$$\Pi(x) = \sum_{i=1}^{3} \lambda_i^{x^i} \cdot \frac{1}{x^i!}$$

To see how this can be used to obtain the mean throughput, note that the conditional throughput, given that the system is at state $x$ is

$$S(x) = x^3 + x^1\sigma = 1 - x^1(1 - \sigma)$$

That is, it is the flow of users (each of which is transmitting a single message) from nodes 1 and 3 through the TDMA channel (see Fig. 5.5).

To find the expected value of this, we require the expected value of $x^1$. This is

$$\begin{aligned} E[x^1] &= \sum_{x \in \mathbf{X}_2 | x^1 = 1} \Pi(x) \\ &= \lambda_1 \cdot \lambda_2 \\ &= 1 - \sigma \end{aligned}$$

The mean throughput, $S$, is thus

$$\begin{aligned} S &= E[S(x)] \\ &= 1 - E[x^1](1 - \sigma) \\ &= 1 - (1 - \sigma)^2 \end{aligned}$$

Other performance measures, such as mean message delay, or probability of buffer overflow, can also be evaluated from the equilibrium distribution, and will be considered at a later stage.

When a product form equilibrium distribution is available, usually an exact analysis of the modelled system is possible. However, many multiple access protocols for communication or computer networks are such that the resulting discrete-time queueing network model has state dependent routing probabilities and, in general, product form solutions are not possible. In such cases, the flow conservation equations are non-linear, and one can either use a numerical solution technique or an approximation method to solve these. Numerical solutions can be prohibitively costly if the state space is large, and we shall mainly be concerned with methods that fall into the approximation category, one of which is described in the following section.

## 5.4 EQUILIBRIUM POINT ANALYSIS

*Equilibrium point analysis* (EPA) is an approximation technique that is applied only to the steady state [TASA 86]. In essence, the equilibrium distribution of the Markov chain to be solved is approximated by a unit impulse located at a point in the state space where the system is in equilibrium (*an equilibrium point*). The method requires the solution of a set of coupled, (usually) non-linear equations, known as the *equilibrium point equations*, which can often be reduced to a fixed point equation for some variable of interest. This, in turn, can be solved by the standard methods for such equations (iteration, bisection).

EPA is tantamount to solving the flow conservation equations (5.39) under the assumptions that *the system is always at an equilibrium point.* Before doing this however, equations (5.39) are written in an equivalent form by first applying Little's result to give

$$E[x^i]\mu_i = \sum_{j=1}^{H} E[x^j]\mu_j r_{ji}, \qquad i = 1, 2, \ldots, H$$

The expected values in the above are then replaced (approximated) by the corresponding point values that solve the equations

$$x^i\mu_i = \sum_{j=1}^{H} x^j\mu_j r_{ji} \qquad i = 1, 2, \ldots, H \tag{5.40}$$

These are the *equilibrium point equations*; the left hand side approximates the expected number of customers leaving node $i$ per unit time, and the right hand side approximates the expected number of customers entering node $i$ per unit time. With a closed network having a fixed population of $N$ customers, one of the equations (5.40) is linearly dependent on the others, and is replaced by the constraint

$$\sum_{i=1}^{H} x^i = N \tag{5.41}$$

Equations (5.40) and (5.41) thus give $H$ independent equations that are usually non-linear, and can be solved to obtain an equilibrium point

$$x_e = (x_e^1, x_e^2, \ldots, x_e^H) \tag{5.42}$$

This is done under the assumption that *the components of the state vector are real valued rather than integer valued quantities.* Note that $x_e$ in equation (5.42) now denotes the equilibrium value of $x$, and the subscript is *not* a time-slot index.

It should be clear from the above that the equilibrium distribution of the Markov chain $\mathbf{x}$ has been approximated by a unit impulse located at $x_e$. Then if $S(x)$ is some performance measure that is a function of $x$, the expected value for this is

$$E[S(x)] = \int_{\mathbf{X}^H} S(x)\delta(x - x_e)\, dx = S(x_e) \tag{5.43}$$

Here, $\mathbf{X}^H$ is a $H$-dimensional Euclidean space that includes the state-space $\mathbf{X}_N$ as a discrete-valued subset. Thus, mean values of performance measures can be approximated by their value at an equilibrium point.

*Example 5.3: Slotted Aloha with delayed first transmission*

To make the idea of EPA clear, we shall apply this to solve the model for slotted Aloha with delayed first transmission that was previously analysed by directly solving the balance equations for the corresponding Markov chain in Example 3.4 [TOBA 80B]. The diagrammatic model was given in Fig. 3.2. This is shown redrawn in Fig. 5.6 with some change in the labelling so that we can interpret the model as a discrete-time queueing network.

In Fig. 5.6, users at node 2 are idle, and have no messages for transmission. They can generate a new message in a slot with probability $\sigma$, in which case they jump to node 1 at the end of the slot. Users that don't generate a message remain at node 2. Users at node 1 thus have a message waiting for transmission. They attempt to transmit this message with probability $p$ (per slot). If only one user attempts a transmission this is successful, and the user jumps to node 2 at the end of the slot. If more than one user attempts a transmission out of the node 1 then a collision occurs, and the relevant users jump back to node 1 at the end of the slot.

In the context of our discrete-time queueing network, the above behaviour implies that users at node 2 are served with probability $\sigma$ (per slot), in which case they jump to node 1 just before the end of the slot. Users at node 1 are served with probability $p$ (per slot). If only one user is served in a slot then this is routed to node 2. If more than one user is served, these are routed back to node 1.

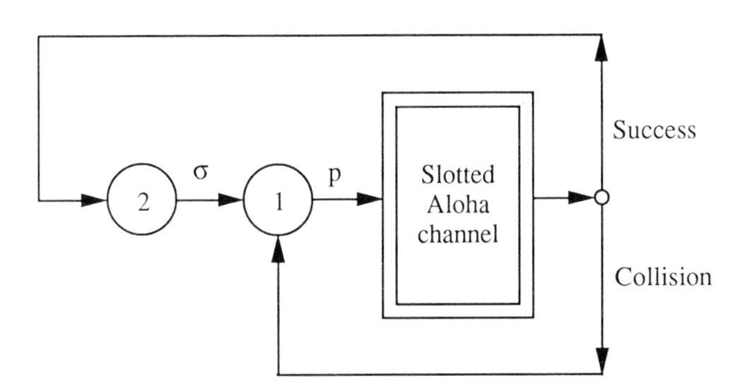


*Figure 5.6 Discrete-time queueing network model for slotted Aloha with delayed first transmission*

The service rates of the nodes are thus

$$\mu_1 = p, \mu_2 = \sigma \tag{5.44}$$

Clearly, the routing probability $r_{21}$, is

$$r_{21} = 1 \tag{5.45}$$

The routing probability $r_{12}$ is just the probability that all but one of the users at node 1 don't attempt a transmission. That is:

$$r_{12} = (1 - p)^{x^1 - 1} \tag{5.46}$$

Then clearly we have (see equation (5.33))

$$r_{11} = 1 - (1 - p)^{x^1 - 1} \tag{5.47}$$

The state of the system is

$$x = x^1 \tag{5.48}$$

since by equation (5.41), knowing $x^1$ implies that we also have $x^2$ as

$$x^2 = N - x^1 \tag{5.49}$$

Note that the routing probabilities $r_{12}$ and $r_{11}$ are no longer independent, but now have a dependency on the state of the system. This statistical dependency arises because of interference between the queues due to the shared broadcast channel. These queueing models are known as interfering queues [KLEI 80], and except in special cases (e.g. [WALR 88]) an exact solution is not possible using classical queueing theory, [TOBA 76], [KLEI 80], [SAAD 81], hence our use of EPA to obtain an approximate solution.

The application of EPA to this two node model is very simple, since there are only two equations involved. That is, for node 2 substitution of (5.44), (5.45), (5.46) and (5.47) into equation (5.40) gives

$$x^2 \sigma = x^1 p(1 - p)^{x^1 - 1} \tag{5.50}$$

The second equation is the constraint (5.41) which gives

$$x^1 + x^2 = N \tag{5.51}$$

The equation for node 1 is not required due to the linear dependence. Solving these equations for $x^1$ we get

$$x^1 = \frac{N\dfrac{\sigma}{p}}{\dfrac{\sigma}{p} + (1 - p)^{x^1 - 1}} \tag{5.52}$$

We thus have a fixed point equation of the form

$$x^1 = f(x^1) \tag{5.53}$$

This can easily be solved for $x^1$ by the standard methods for such equations, such as simple iteration or bisection [PRES 92].

Let a solution to this equation be $x_e^1$; thus $x_e^1$ is an equilibrium point. The conditional throughput, given that the system is at state $x^1$ is clearly

$$S(x^1) = x^1 p(1 - p)^{x^1 - 1} \qquad (5.54)$$

The mean value of this according to equation (5.43) is thus

$$E[S(x^1)] \simeq S(x_e^1) = x_e^1 p(1 - p)^{x_e^1} \qquad (5.55)$$

Note the simplicity of this approach in comparison with using a direct solution of the balance equations of the Markov chain, which was done in Example 3.4. The key thing here is that EPA avoids having to compute the one-step transition probabilities of the Markov chain. Because of this, the method can be used to obtain approximate solutions to quite complex models which, in the absence of a product form distribution, would otherwise be intractable.

To give an example of EPA applied to a multidimensional Markov chain consider the following.

*Example 5.4: Slotted Aloha with delayed first transmission and R-slot channel delay.*

This example is the same as Example 5.3, but with an $R$-slot delay on the channel. The discrete-time queueing network model for this is shown in Fig. 5.7, which is identical to Fig. 5.6 but with the addition of $2R$ nodes. Each of these nodes has a service rate of 1, and acts as a unit delay for any users passing through. This means that users involved in either successful transmissions or collisions have to wait $R$ slots before they return to the idle node, which is labelled $2R + 1$, or node 0.

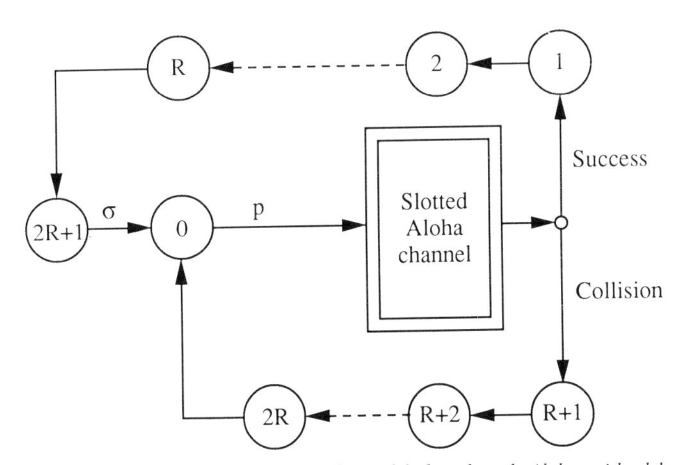


*Figure 5.7 Discrete-time queueing network model for slotted Aloha with delayed first transmission and an R-slot channel delay*

With reference to Fig. 5.7, let $x_n^i$ be the number of users at node $i$ during slot $n$, $i = 0, 1, \ldots, 2R + 1$. Then

$$x_n = (x_n^0, x_n^1, \ldots, x_n^{2R}) \tag{5.56}$$

is a state vector of the Markov chain

$$\mathbf{x} = \{x_n = x, n = 0, 1, 2, \ldots\} \tag{5.57}$$

where

$$x = (x^0, x^1, \ldots, x^{2R}) \in \mathbf{X}_N \tag{5.58}$$

and

$$\mathbf{X}_N = \left\{ x \in \mathbf{X} \,\middle|\, \sum_{i=0}^{2R+1} x^i = N \right\} \tag{5.59}$$

$$\mathbf{X} = \{0, 1, 2, \ldots\}^{2R+1} \tag{5.60}$$

The service rates for the nodes are

$$\mu_0 = p, \mu_1 = \mu_2 = \ldots = \mu_{2R} = 1, \mu_{2R+1} = \sigma \tag{5.61}$$

As for Example 5.3 the routing probabilities are

$$r_{01} = (1 - p)^{x^0 - 1} \tag{5.62}$$

$$r_{0,R+1} = 1 - (1 - p)^{x^0 - 1} \tag{5.63}$$

In addition, we have

$$r_{12} = r_{23} = \ldots = r_{R-1,R} = r_{R,2R+1} = r_{2R+1,0} = 1$$

$$r_{R+1,R+2} = r_{R+2,R+3} = \ldots = r_{2R-1,2R} = r_{2R,0} = 1 \tag{5.64}$$

Equations (5.40) become

$$x^{2R+1}\sigma = x^R \tag{5.65}$$

$$x^R = x^{R-1} = \ldots = x^1 = x^0 p(1 - p)^{x^0 - 1} \tag{5.66}$$

$$x^{2R} = x^{2R-1} = \ldots = x^{R+1} = x^0 p([1 - (1 - p)^{x^0 - 1}] \tag{5.67}$$

In addition, we have equation (5.41) as

$$\sum_{i=0}^{2R+1} x^i = N \tag{5.68}$$

It is unnecessary to write down the equation for node 0 since it is linearly dependent on the others. Again it is worth mentioning that in the equilibrium point equations (5.65), (5.66) and (5.67) above, the left hand side represents the expected number of users leaving the corresponding node per unit time, and the right hand side the expected number of users entering the node per unit time.

Substituting from equations (5.66), (5.67) and (5.68) into (5.65) then

$$(N - x^0 - Rx^0 p)\sigma = x^0 p(1 - p)^{x^0 - 1} \tag{5.69}$$

Manipulating this we get

$$x^0 = \frac{N\frac{\sigma}{p}}{R\sigma + \frac{\sigma}{p} + (1 - p)^{x^0 - 1}} \tag{5.70}$$

Note that by setting $R = 0$ in equation (5.70), and writing $x^0 = x^1$, we get equation (5.52) as we should expect (zero delay on the channel).

Throughput can be obtained in the same way as in Example 5.3. That is, solve the fixed point equation (5.70) to obtain an equilibrium point value $x_e^0$, and then use this to compute the mean throughput as

$$S = S(x_e^0) = x_e^0 p(1 - p)^{x_e^0 - 1} \tag{5.71}$$

Comparing Examples 5.3 and 5.4 we note that using EPA the only added complexity when solving a multidimensional Markov chain as opposed to a single dimensional one is in solving the set of coupled non-linear equations and reducing these to a single fixed point equation. The computational cost of solving the latter does not increase, and is often negligible.

Once throughput has been calculated, the other main performance measure, the mean message delay can be obtained using the throughput and Little's result. Since the mean message delay, $W$, is defined as the mean time from when a message is generated to the successful completion of its transmission, with reference to Fig. 5.7 this is therefore the time from when a user at node $2R + 1$ generates a message to the time of the user's return to this node.

Denoting the mean number of messages in the system by $\bar{n}$, then by Little's result we have

$$\begin{aligned} W &= \frac{\bar{n}}{S} \\ &= \frac{N - x^{2R+1}}{S} + \frac{x^{2R+1}\sigma}{S} \end{aligned} \tag{5.72}$$

Applying Little's result to node $2R + 1$ we have

$$x^{2R+1} = S\frac{1}{\sigma} \tag{5.73}$$

From equations (5.72) and (5.73) we get

$$W = \frac{N}{S} - \frac{1}{\sigma} + 1 \tag{5.74}$$

Since $N$ and $\sigma$ are known system parameters, then once $S$ is found, $W$ can be obtained directly from this using (5.74).

Note that this same formula will also give the mean message delay for the model with zero channel delay considered in Example 5.3, since the effect of the channel delay is implicitly contained in the value of $S$.

## 5.5 DIFFERENT CUSTOMER CLASSES

So far we have considered discrete-time queueing networks where the customers all have identical statistical properties (service time requirements, routing probabilities). In practice, this is not necessarily the case, and some customers may have different requirements to others. All customers that have statistically identical requirements are said to belong to the same *customer class.*

For the case of a closed network with $N$ customers, then the number of possible customer classes, $c$, is clearly in the range $1 \leq c \leq N$. The networks considered so far are thus special cases where $c = 1$. In what follows we develop extensions to take into account different customer classes. We shall confine attention to closed networks of discrete-time $M^{a_n}/M^{d_n}/\infty$ nodes as used in the discrete-time queueing models for communication or computer networks. The routing probabilities will be assumed state dependent, and EPA will be used to obtain a solution. Two methods will be considered, one of which is based on a recursion, the other on iteration.

### 5.5.1 Recursive EPA

Because the equilibrium point equations are, in general, non-linear, a solution to the EPA equations usually reduces to solving a fixed-point equation of the form

$$x^j = f(x^j) \tag{5.75}$$

where $x^j$ is a component of the state vector in terms of which all the other components have been found. This approach requires all the network's customers to belong to the same class and is the standard form of EPA.

Using the notation $x(K)$ to denote the value of $x^j$ when the network is populated by $K$ customers, an alternative approach is to obtain a function $\phi[x(K)]$ such that

$$x(K) = f(\phi[x(K-1)]) \tag{5.76}$$

where

$$\phi[x(0)] \equiv 0 \tag{5.77}$$

If such a function can be found, the fixed point equation (5.75) can be replaced by the recursion (5.76) which can then be evaluated up to the population level of interest.

We shall consider the case of $N$ customer classes, since the generalisation from this will be obvious and will be discussed later. Then define $x_{1,2,\ldots,K}(K)$ to be the value of $x^j$ for a network consisting of a $K$-customer subset $\{1, 2, \ldots, K\}$, $K \leq N$, of the total network population. Also let $x^{(i)}_{1,2,\ldots,K}(K)$ be the conditional probability that customer $i$ is at node $j$, $i \in \{1, 2, \ldots, K\}$, given that the network is at state $x$. If customers are assumed to behave independently, then

$$x_{1,2,\ldots,K}(K) = \sum_{i=1}^{K} x^{(i)}_{1,2,\ldots,K}(K) \tag{5.78}$$

Likewise, any performance measure $S(x^i)$ that is a linear function of $x^j = x_{1,2,\ldots,K}(K)$ can be obtained as

$$S(x^j) = \sum_{i=1}^{K} S(x^{(i)}_{1,2,\ldots,K}(K)) \tag{5.79}$$

Now define, $\phi[.]$ to be a function of the state dependent routing probabilities such that

$$x^{(i)}_{1,2,\ldots,K}(K) = f(\phi'_1(K-1)) \tag{5.80}$$

where

$$\phi'_i(K-1) = \phi[x_{1,2,\ldots,i-1,i+1,\ldots,K}(K-1)] \tag{5.81}$$

and $f(.)$ is a function that results from a solution to the equilibrium point equations.

The interpretation of equation (5.80) is that in a network with $K$ customers in equilibrium, at the jumps a customer sees routing probabilities that are a function of the equilibrium point of a network populated by the other $K-1$ customers. This is an intuitive assertion motivated by the Arrival Theorem for continuous time networks [LAVE 80], [SEVC 81]. This latter result applies to continuous-time networks of quasi-reversible queues with independent routing probabilities, and basically states that a customer that jumps in such networks sees the other customers with their equilibrium distribution when that customer is not in the network. A good discussion of this can be found in [WALR 88]. The result will not be proved here, but will be used in proving the following.

*Equilibrium Point Arrival Theorem:* For any queueing network in equilibrium for which the Arrival Theorem holds, then with the network constrained to operate only at an equilibrium point, a customer that jumps sees the others at their equilibrium point.

*Proof:* Applying the conditions of the theorem, with the network operating only at an equilibrium point then the equilibrium distribution becomes a unit impulse located at the equilibrium point. Since the Arrival Theorem holds, the result follows immediately.

Clearly, the use of the above theorem in the context of the present networks is essentially an approximation. The accuracy of this will be considered later when numerical results for such networks are discussed.

The use of the Equilibrium Point Arrival Theorem implies that the scenario at the jumps is as follows. With the network in equilibrium, a customer $i$ that jumps will, because of the state dependent routing probabilities, be routed to the next node with a probability that is dependent on the equilibrium point of the other customers in the network. This routing probability will be fixed at a value given by a solution to the equilibrium point equations for a network with customer $i$ removed.

Having solved the equilibrium point equations to obtain a function in the form of equation (5.80), a procedure for calculating the performance measures of the individual customers when these belong to different customer classes is as follows.

*Step 0:* $x(0) \equiv 0,\ \phi_i'(0) \equiv 0, \qquad i = 1, 2, \ldots, N$

*Step 1:* Calculate $x^{(i)}(1),\ \phi[x_i(1)], \qquad i = 1, 2, \ldots, N$

*Step 2:* Calculate
$$x_{i,j}^{(i)}(2) = f(\phi_i'(1))$$
$$x_{i,j}^{(j)}(2) = f(x_j'(1))$$
$$\phi[x_{i,j}(2)]$$
for all $\binom{N}{2}$ combinations of $i,j \in \{1, 2, \ldots, N\}$

*Step 3:* Calculate
$$x_{i,j,k}^{(i)}(3) = f(x\phi_i'(2))$$
$$x_{i,j,k}^{(j)}(3) = f(\phi_j'(2))$$
$$x_{i,j,k}^{(k)}(3) = f(\phi_k'(2))$$
$$\phi[x_{i,j,k}(3)]$$
for all $\binom{N}{3}$ combinations of $i, j, k \in \{1, 2, \ldots, N\}$

$\vdots$

*Step N:* Calculate $x_{1,2,\ldots,N}^{(i)}(N) = f(\phi_i'(N-1))$
for all $i = 1, 2, \ldots, N$.

Performance measures can be obtained from the results of Step $N$ above, as shown in equation (5.79).

A more concise statement of the above algorithm in the form of pseudocode is as follows:

```
Set x(0) = φ'_i(0) = 0, for i = 1 to N
K := 1
While K ≤ N do
    for i = i1 to iK do
        Calculate x^(i)_{i1,i2,...,iK}(K) = f(φ'_i(K − 1))
```

and $\phi[x_{i1,i2,\ldots,iK}(K)]$
for all ${}^{N}C_{K}$ combinations of $i1, i2, \ldots, iK \in \{1, 2, \ldots, N\}$
$K := K + 1$

As an example of applying the above, consider the following.

*Example 5.5: Slotted Aloha with delayed first transmission and different customer classes*

The specification is identical with that of Example 5.3, except we shall consider each user belonging to a different class.

The multiclass queueing network model for $N$ user classes is shown in Fig. 5.8, where nodes 1 and 2 have their previous meanings (as in Example 5.3). The difference is that each user $i$ can now generate a message node 2 with probability $\sigma_i$, $i = 1, 2, \ldots, N$, and messages are transmitted/retransmitted from node 1 with probability $p_i$, $i = 1, 2, \ldots, N$, where $\sigma_i$ and $p_i$ can be different for each $i$.

With the network populated by $K$ users, $1 \leq K \leq N$, a user $i$ at node 2 is served in a slot with probability $\sigma_i$ and thus jumps from the node just before the end of the slot with this probability, and is routed to node 1. Similarly, a user $i$ is served at node 1 with probability $p_i$ and jumps from the node just before the end of the slot with this probability. Routing from this node is state dependent, so applying the Equilibrium Point Arrival Theorem, at the jump this user sees the other $K - 1$ users at their equilibrium point. Then the probability of at least one transmission/retransmission attempt from node 1 in a network populated by the other $K - 1$ users will give the probability that user $i$ is involved in a collision, and so is routed back to node 1. Defining this probability as $\phi'_i(K - 1)$, then

$$r_{11} = \phi'_i(K - 1) \tag{5.82}$$

$$r_{12} = 1 - \phi'_i(K - 1) \tag{5.83}$$

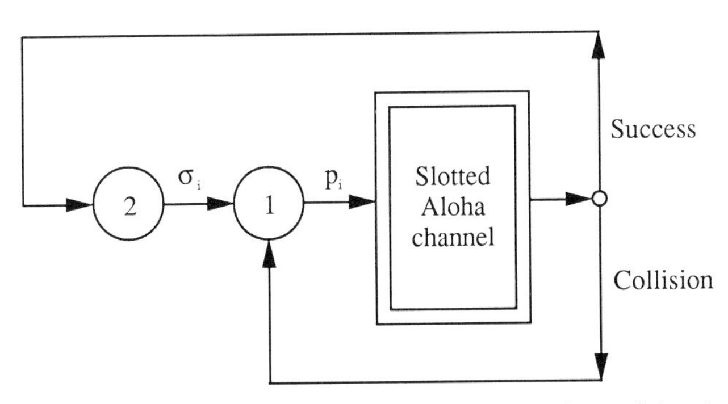


*Figure 5.8 Discrete-time queueing network model for customer class i of slotted Aloha with delayed first transmission*

At node 2 we have, as before

$$r_{21} = 1 \tag{5.84}$$

Defining $x_{1,2,\ldots,K}(K)$ to be the expected number of users at node 1 when there are $K$ users in the network, then the equilibrium point equations for user $i$ at node 2 are

$$(1 - x^{(i)}_{1,2,\ldots,K}(K))\sigma_i = x^{(i)}_{1,2,\ldots,K}(K)p_i(1 - \phi'_i(K-1)) \qquad i = 1, 2, \ldots, N \tag{5.85}$$

Note that in the above, the left hand side of the equation (5.85) represents the expected number of users *in class i* moving out of node 2 per unit time, and the right hand side represents the expected number of users *in class i* entering node 2 per unit time (in this case user $i$ and class $i$ are the same thing in that we are considering only one user per class).

Rearranging equation (5.85) we get

$$x^{(i)}_{1,2,\ldots,K}(K) = \frac{\sigma_i/p_i}{\sigma_i/p_1 + (1 - \phi'_i(K-1))} \tag{5.86}$$

Since users are assumed to behave independently, then

$$\phi[x_{1,2,\ldots,K}(K)] = 1 - \prod_{i=1}^{K} (1 - x^{(i)}_{1,2,\ldots,K}(K)p_i) \tag{5.87}$$

Equation (5.87) gives the probability that one or more users attempt to transmit in a network of $K$ users. Then $\phi'_i(K-1)$ can be obtained from equation (5.87) using equation (5.81).

Equations (5.86) and (5.87) can now be solved for increasing population levels using the procedure previously given. The throughput of user $i$ is then

$$S(x^{(i)}_{1,2,\ldots,N}(N)) = x^{(i)}_{1,2,\ldots,N}(N)p_i(1 - \phi'_i(N-1)) \qquad i = 1, 2, \ldots, N \tag{5.88}$$

The total mean throughput of the network is

$$S(x_{1,2,\ldots,N}(N)) = \sum_{i=1}^{N} S(x^{(i)}_{1,2,\ldots,N}(N)) \tag{5.89}$$

Delay for user $i$, $W(x^{(i)}_{1,2,\ldots,N}(N))$, is obtained using Little's result (as for the network with one customer class, see equation (5.74)) as

$$W(x^{(i)}_{1,2,\ldots,N}(N)) = \frac{1}{S(x^{(i)}_{1,2,\ldots,N}(N))} - \frac{1}{\sigma_i} + 1 \qquad i = 1, 2, \ldots, N \tag{5.90}$$

Mean message delay for the network is thus

$$W(x_{1,2,\ldots,N}(N)) = \frac{1}{N} \sum_{i=1}^{N} W(x^{(i)}_{1,2,\ldots,N}(N)) \tag{5.91}$$

Note the parallel between the procedure adopted here and that for a single customer class, as in Example 5.3. In particular, compare equations (5.86) with (5.52) and (5.88) with (5.54).

Although the procedure and example given above are for the case where each user belongs to a different class, the procedure generalises in a rather obvious way to the case of $c$ customer classes, $1 \leq c \leq N$. If at a particular population level, for a given subset of customers there is more than one customer in the same class, then a calculation need only be carried out once for the class, since the results will be identical for all customers in the same class.

When $c = 1$ then time complexity of the algorithm will be of order $O(N)$, where $O(N)$ is a function of $N$ such that $O(N)/N$ converges to a positive number when $N \to \infty$. In the worst case when $c = N$, due to the combinatorial nature of the algorithm, time complexity will be of order $O(2^N - 1)$. In general, if $N_r$ is the number of customers in class $r$, time complexity is of order $O(\Pi_{r=1}^{C}(N_r + 1) - 1))$. Note that unlike algorithms such as mean value analysis (MVA) [REIS 80], this complexity is independent of the number of nodes. Then provided the number of customer classes and users remains the same, the time complexity will not increase with the degree of complication of a protocol, since the latter will just add to the number of nodes in the queueing model. The trade-off here is a more difficult solution to the EPA equations. Further discussions of this method can be found in [WOOD 93A] and [WOOD 93B].

### 5.5.2 Extended EPA

As an alternative to the recursive method considered in the previous sub-section, an iterative form of the method can be developed as follows.

Again consider a network with $N$ customers, each belonging to a different class, and let $x^j$ in equation (5.75) be denoted simply as $x$. Furthermore, denote by $x_i$ the conditional probability that user $i$ is at node $j$, given the network state. Then if customers are assumed to behave independently

$$x^j = x = \sum_{i=1}^{N} x_i \tag{5.92}$$

Likewise, any performance measure that is a linear function of $x$, say $S(x)$, can be obtained as

$$S(x) = \sum_{i=1}^{N} S(x_i) \tag{5.93}$$

Now again define the function $\phi[.]$ to be a function of the state dependent routing probabilities such that

$$x_i = f(\phi'(x_i)) \tag{5.94}$$

where

$$\phi'(x_i) = \phi[x_1, x_2, \ldots, x_{i-1}, x_{i+1}, \ldots, x_N] \tag{5.95}$$

and $f(.)$ is a function that results from a solution to the EPA equations.

Equation (5.94) is again motivated by the Arrival Theorem as in the recursive form of EPA. The idea is then to proceed, as before, and solve the EPA equations for a single user on the basis that any state dependent routing probabilities are functions of the equilibrium point of the other $N-1$ users. That is, assume the Equilibrium Point Arrival Theorem holds. A second key assumption is however now invoked, that *the fraction of users at particular nodes is a linear function of the network population.* This is a similar approximation to that used in some approximate mean value analysis (MVA) algorithms, such as that known as Linearizer [CHAN 82].

Having found an expression in the form of equation (5.94), this can then be used in the following iterative algorithm, which takes into account the above linearising approximation.

Let $x_i(n)$ be the value of $x_i$ after $n$ iterations, and let $\Delta$ be an acceptable error tolerance.

$n := 0$
For $i := 1$ to $N$ do
  $x_i(n) := 1/H$ {$H$ is the total number of nodes}
Repeat
  $n := n + 1$
  For $i := 1$ to $N$ do
    $\phi'(x_i(n)) := \phi(x_2(n-1), \ldots, (x_N(n-1));$  $i = 1$
      $\phi(x_1(n), \ldots, x_{i-1}(n), x_{i+1}(n-1), \ldots, x_N(n-1));$  $2 \leq i \leq N-1$
      $\phi(x_1(n), \ldots, x_{N-1}(n));$  $i = N$
    $x_i(n) := f(\phi'(x_i(n)));$
Until $(|x_i(n) - x_i(n-1)| \leq \Delta$ for all $i \in \{1, 2, \ldots, N\});$

Assuming the algorithm terminates after $k$ iterations, then because of the independence assumption for users, performance measures can be calculated using equation (5.93) with $x_i = x_i(k)$.

Note that the above is a form of Gauss-Seidel iteration that uses the most up to date estimate in its calculations. The algorithm exhibits usual convergence properties for such multivariate iterations. See, for example, [ORTE 70].

An example follows.

*Example 5.6: Slotted Aloha with delayed first transmission and different customer classes*

This is identical to Example 5.5 in order to illustrate the similarities between the recursive and iterative forms of EPA for different customer classes.

Defining $x_i$ to be the number of users at node 1, and $\phi'(x_i)$ to be the probability of at least one transmission attempt from a user at node 1 in a network populated by the $N-1$ users $1, 2, \ldots, i-1, i+1, \ldots, N$, then using the same arguments as in Example 5.5 we have

$$r_{11} = \phi'(x_i) \tag{5.96}$$

$$r_{12} = 1 - \phi'(x_i) \tag{5.97}$$

$$r_{21} = 1 \tag{5.98}$$

Then the equilibrium point equation for user $i$ at node 2 is

$$(1 - x_i)\sigma_i = x_i p_1 (1 - \phi'(x_i)) \tag{5.99}$$

Rearranging this we get

$$x_i = \frac{\sigma_i/p_i}{\sigma_i/p_i + (1 - \phi'(x_i))} \tag{5.100}$$

Since users are assumed to behave independently, then $\phi'(x_i)$ is given by

$$\phi'(x_i) = 1 - \prod_{\substack{j=1 \\ j \neq i}}^{N} (1 - x_j p_j) \tag{5.101}$$

Equations (5.100) and (5.101) can now be solved for $x_i$ using the iterative algorithm specified. Let $x_{ie}$ be a solution. The mean throughput of user $i$ is then

$$S(x_{ie}) = (1 - x_{ie})\sigma_i \tag{5.102}$$

The total mean throughput of the network is

$$S(x_e) = \sum_{i=1}^{N} S(x_{ie}) \tag{5.103}$$

Mean message delay for user $i$, $W(x_{ie})$, is

$$W(x_{ie}) = \frac{1}{S(x_{ie})} - \frac{1}{\sigma_i} + 1 \tag{5.104}$$

Mean message delay for the network is

$$W(x_e) = \frac{1}{N} \sum_{i=1}^{N} W(x_{ie}) \tag{5.105}$$

Clearly, there are similarities between the iterative and recursive methods. The former however has a time complexity of order $O(N)$, which is independent of the number of customer classes.

In subsequent references, the recursive form of EPA will be abbreviated as REPA, and the iterative form as EEPA (Extended EPA), since this latter method is, in effect, a multivariate extension of the standard form of EPA.

## 5.6 EXERCISES

5.1 Consider two queues of the discrete-time $M^{a_n}/M^{d_n}/\infty$ type, where the output of queue 1 is connected to the input of queue 2, and the output of queue 2 is connected back to the input of queue 1. No other feedback is involved. Use time reversal arguments (Kelly's lemma) to show that this arrangement has a product form equilibrium distribution.

5.2 Consider two discrete-time $M/M/1$ queues in tandem. The input to queue 1 is a Bernoulli stream with parameter $p$, and the output of queue 1 is connected directly to queue 2's input. Show that the output from queue 2 is also a Bernoulli stream with parameter $p$.

5.3 If in the arrangement of Exercise 5.2 an additional Bernoulli stream is connected to the input of queue 2, show that the system cannot have a product form equilibrium distribution. Hint: Show that the Bernoulli distribution does not have the superposition property (unlike the Poisson distribution).

5.4 Modify the slotted Aloha model with delayed first transmission, as specified in Example 5.3, so that each user has a buffer that can hold two messages. Solve this three node model using EPA and hence find expressions for the throughput and mean message delay.

5.5 Repeat Exercise 5.4 for the case of $N$ customer classes, with a single user in each class. Assume that REPA is to be used to obtain a solution.

# 6. SATELLITE NETWORKS

*Chapter Objectives: To demonstrate the application of various modelling concepts, using satellite networks as examples. The modelling objectives are summarised at the start of each section.*

So far we have mainly been concerned with general issues pertaining to Markov chains, discrete-time queues, and discrete-time queueing networks. The orientation is now shifted more towards application issues in the area of communication and computer networks, where we make use of the theoretical concepts so far considered.

The first type of communication network we shall examine is the satellite network. The single major feature that characterises a satellite network is the propagation delay on the channel. For example, for a geostationary satellite orbiting the earth at an altitude of 3600 km, although the signals propagate at the speed of light, the round-trip delay from an earth station out to the satellite and back is of the order of 0.27 seconds. This is something of a megatime in the context of a digital transmission system, and the delay is physically apparent to a user during telephone conversations, or other real-time interactions via satellite, such as live television interviews.

In modelling a satellite network the influence of the channel propagation delay must be explicitly taken into account, and this can give rise to certain complications in the model, such as a multidimensional sate-space. In all the models in this chapter, the following common assumptions will be made.

A6.1: Each data packet is of fixed length $L$ bits.

A6.2: A satellite broadcast channel with (data) bit-rate $C$ bits/s is used.

A6.3: Channel time is divided into slots whose duration is equal to the transmission time of a packet; that is, $L/C$ seconds.

A6.4: The round trip propagation delay of the channel is equal to $R$ slots between any two earth stations.

A6.5: The channel is error-free except for collisions.

A6.6: The network has a fixed population of $N$ users.

A6.7: The interarrival time (in slots) between messages for a user $i$ is geometrically distributed with mean $1/\sigma_i$; that is, a user generates a message with probability $\sigma_i$ per slot, where $\sigma_i \ll 1$. With only one user class, then $\sigma_i = \sigma$, $i = 1, 2, \ldots, N$.

A6.8: Each message consists of a number of packets, and the message length is geometrically distributed with mean $1/\gamma_i$ packets for user $i$. With only one user class, then $\gamma_i = \gamma$, $i = 1, 2, \ldots, N$.

Additional assumptions will be introduced as necessary for specific protocols.

A unified approach is adopted in modelling all the protocols considered by first developing discrete-time queueing network models for the protocols, and then solving these using either the product form (where applicable) or EPA. To this end Section 6.1 considers fixed assignment, time division multiple access (TDMA) protocols, while Section 6.2 covers the random access, slotted Aloha type protocols. Section 6.3 generalises the previous results by developing a code division multiple access (CDMA) model, from which it is apparent that TDMA and slotted Aloha are special cases. Section 6.4 considers the case of buffered slotted Aloha, where each user supports a buffer of finite size for queueing packets awaiting transmission. Each of these models has been chosen to illustrate a particular modelling concept or technique. This objective is summarised at the start of each section.

Performance results for both throughput and mean message (or packet) delay are presented for each of the models in the appropriate sections. Since the main focus of the text is on the modelling techniques rather than the protocols, the results are representative samples only, and are not intended as an in depth study of the protocols. Their main function is to give an assessment as to the accuracy of the modelling methods used, when compared with simulations.

## 6.1 TIME DIVISION MULTIPLE ACCESS

*Modelling objectives: To show how fixed assignment protocols result in product form equilibrium distributions, and see how these can be used in performance evaluation.*

*Time division multiple access* (TDMA) is a fixed assignment protocol in which channel time is divided into *frames*, where each frame consists of $N$ slots. Each user $i$ is assigned a slot $i$, $i = 1, 2, \ldots, N$, in each frame. In this way each user is guaranteed a bandwidth of $C/N$ bits/s, *independent* of the other users.

A single user class, discrete-time queueing network model for TDMA is shown in Fig. 6.1, where assumptions A6.1–A6.8 apply. This is a generalisation of Example 5.2 to the case of $N$ users and geometrically distributed message lengths. Users at nodes $1 \leq i \leq N$ have no messages waiting but generate a message at the start of a slot with probability $\sigma$, or generate no message with probability $1 - \sigma$. Users at nodes $N + 1 \leq i \leq 2N$ have a message waiting. A user whose slot position will appear in a frame after $i$ slots is at either node $i + 1$ or at node $i + N + 1$, $0 \leq i \leq N - 1$.

Then a user at node 1 or $N + 1$ can transmit a packet in the current slot. Note that the number of users at each node is either 0 or 1, and the

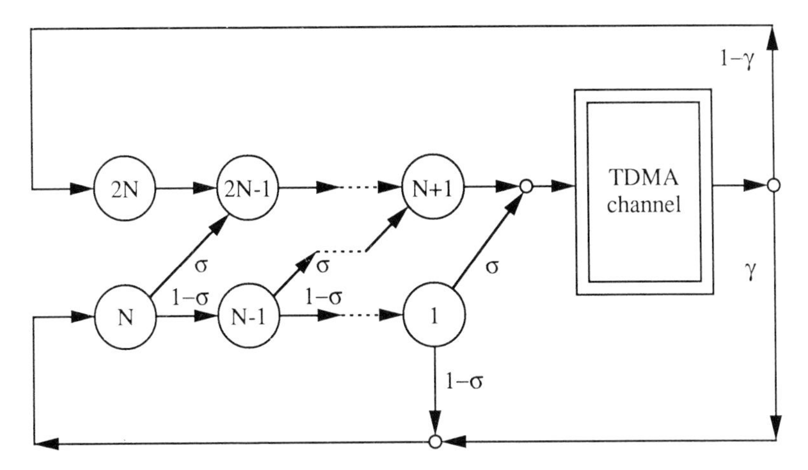


*Figure 6.1 Discrete-time queueing network model for TDMA with multipacket messages*

sum of users at nodes $i$ and $i + N$, $i = 1, 2, \ldots, N$, is always equal to 1. If a user at node $i$, $2 \leq i \leq N$, generates a message, the user jumps to node $i + N - 1$ at the end of the slot, otherwise the user jumps to node $i - 1$. A user that generates a message at node 1 or is at node $N + 1$ transmits a packet and then jumps to either node $2N$ or $N$ according to whether the message contains more packets or not, respectively. Note that these users pass through a block labelled 'TDMA channel', which implies these users transmit their packets.

The state vector of this model is

$$x = (x^1, x^2, \ldots, x^N) \tag{6.1}$$

The state space is

$$\mathbf{X}_N = \{x \in \mathbf{X} | x^i \in \{0, 1\}, x^i + x^{i+N} = 1, \qquad i = 1, 2, \ldots, N\} \tag{6.2}$$

$$\mathbf{X} = \{0, 1, 2, \ldots\}^N \tag{6.3}$$

The service rates at the nodes are

$$\mu_1 = \mu_2 = \cdots = \mu_{2N} = 1 \tag{6.4}$$

The non-zero routing probabilities are

$$r_{i,i-1}, N + 2 \leq i \leq 2N \tag{6.5}$$

$$\left.\begin{aligned} r_{i,i+N-1} &= \sigma \\ r_{i,i-1} &= 1 - \sigma \end{aligned}\right\} 2 \leq i \leq N \tag{6.6}$$

$$r_{N+1,2N} = 1 - \gamma \tag{6.7}$$

$$r_{N+1,N} = \gamma \tag{6.8}$$

$$r_{1,2N} = \sigma(1-\gamma) \tag{6.9}$$

$$r_{1,N} = 1 - \sigma(1-\gamma) \tag{6.10}$$

The routing probabilities are clearly independent, and so the model will have a product form equilibrium distribution, which is given by equation (5.35) as

$$\Pi(x) = G_N \prod_{i=1}^{2N} \left(\frac{\lambda_i}{\mu_i}\right)^{x^i} \frac{1}{x^i!} \tag{6.11}$$

By normalisation

$$\sum_{x \in \mathbf{X}_N} \Pi(x) = G_N \sum_{x \in \mathbf{X}_N} \prod_{i=1}^{2N} \left(\frac{\lambda_i}{\mu_i}\right)^{x^i} \frac{1}{x^i!} = 1 \tag{6.12}$$

The $\lambda_i$ are obtained from a solution to the flow conservation equations, and the state space constraint equations

$$\lambda_i + \lambda_{i+N} = 1, \qquad i = 1, 2, \ldots, N \tag{6.13}$$

We thus have

$$\left.\begin{aligned} \lambda_1 &= \lambda_2(1-\sigma) \\ \lambda_2 &= \lambda_3(1-\sigma) \\ &\vdots \\ \lambda_{N-1} &= \lambda_N(1-\sigma) \end{aligned}\right\} \tag{6.14}$$

$$\lambda_N = \lambda_1(1 - \sigma(1-\gamma)) + \lambda_{N+1}\gamma \tag{6.15}$$

Then from equations (6.14)

$$\lambda_2 = \frac{\lambda_1}{1-\sigma}$$

$$\lambda_3 = \frac{\lambda_1}{(1-\sigma)^2}$$

$$\vdots$$

$$\lambda_N = \frac{\lambda_1}{(1-\sigma)^{N-1}} \tag{6.16}$$

Also, from equations (6.13) and (6.15)

$$\lambda_N = \lambda_1(1-\gamma)(1-\sigma) + \gamma \tag{6.17}$$

Solving for $\lambda_1$ from (6.16) and (6.17)

$$\lambda_1 = \frac{\gamma(1-\sigma)^{N-1}}{1-(1-\gamma)(1-\sigma)^N} \tag{6.18}$$

From equations (6.16) and (6.18) we thus have a general solution to the flow conservation equations as

$$\lambda_i = \frac{\gamma(1-\sigma)^{N-i}}{1-(1-\gamma)(1-\sigma)^N} \tag{6.19}$$

and

$$\lambda_{i+N} = 1 - \lambda_i \qquad i = 1, 2, \ldots, N \tag{6.20}$$

Having found the $\mu_i$ and $\lambda_i$ one could proceed from this point and evaluate the normalising constant $G_N$ using equation (6.12). This, however, is unnecessary if we are only interested in mean values of performance measures. For example, the conditional throughput, given that the system is at state $x$ is

$$S(x) = x^{N+1} + x^1(1-\sigma) \tag{6.21}$$

Using the restriction on the state space, this becomes

$$S(x) = 1 - x^1(1-\sigma) \tag{6.22}$$

The expected value of throughput is thus

$$E[S(x)] = S = 1 - E[x^1](1-\sigma) \tag{6.23}$$

Since by Little's result

$$E[X^1] = \frac{\lambda_1}{\mu_1} = \lambda_1 \tag{6.24}$$

equations (6.18), (6.23) and (6.24) give

$$S = \frac{1-(1-\sigma)^N}{1-(1-\sigma)^N(1-\gamma)} \tag{6.25}$$

This is in units of messages per mean message transmission time. In units of messages per slot, the throughput becomes $S\gamma$.

To calculate delay, we again use Little's result. If the expected number of users at nodes $N+1, N+2, \ldots, 2N$ is $\bar{n}$, then the mean message access delay (queueing delay) is

$$Q = \frac{\bar{n} + (N-\bar{n})\sigma}{S\gamma} \qquad \textit{slots} \tag{6.26}$$

Since we have

$$(N-\bar{n})\sigma = S\gamma \tag{6.27}$$

then delay becomes

$$Q = \frac{N}{\gamma S} - \frac{1}{\sigma} + 1 \tag{6.28}$$

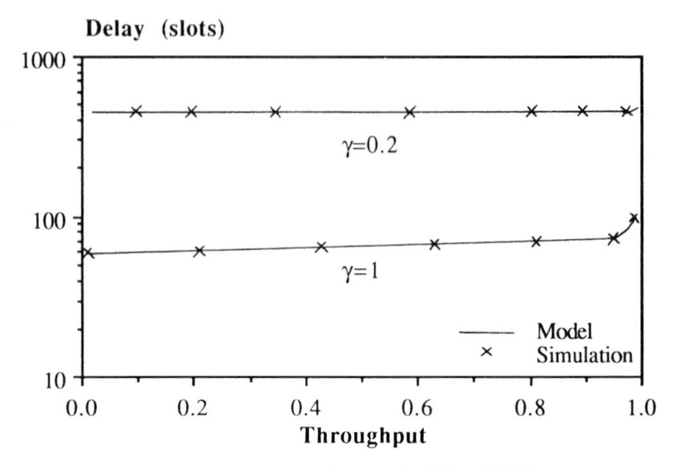


*Figure 6.2 Delay-throughput tradeoff for TDMA*

The mean message delay, $W$, assuming a channel propagation delay of $R$ slots is therefore

$$W = \frac{N}{\gamma S} - \frac{1}{\sigma} + R + 1 \tag{6.29}$$

The mean message delay and throughput tradeoff for a system with $N = 100$, $R = 12$ and different message lengths is shown plotted in Fig. 6.2. The comparison is with simulation results taken over a time span of $10^5$ slots, and it can be seen that for all practical purposes the model gives exact results.

An important point of note here is that the flow conservation equations can be solved exactly in this case. This is because the time delay through a node $i$ is simply $1/\mu_i$, which is independent of the number of users at node $i$, plus the fact that the routing probabilities are independent. Then equations (6.13), (6.14) and (6.15) give a set of $N$ independent, linear equations which can be solved for the $\lambda_i$. Really this is just EPA in disguise, which, in this particular case would give an identical exact result. The application of EPA to a model of this type can be found in Tasaka's book [TASA 86]. Other approaches to modelling TDMA can be found in [HAYE 74], [LAM 76], [KOSO 78], [HILL 87] and [ROM 90].

## 6.2 SLOTTED ALOHA

### 6.2.1 Zero channel delay

*Modelling objectives: To show how EPA can be used to evaluate the performance of a simple model with state dependent routing probabilities.*

In this section we consider the basic model for the well known slotted Aloha protocol with zero channel propagation delay. This has been analysed previously with a delayed first transmission assumption. Here this assumption will be relaxed. Assumptions A6.1 to A6.8 apply, along with the additional assumptions as follows:

A6.2.1: There is a single customer class.
A6.2.2: Messages consist of single packets only ($\gamma = 1$).
A6.2.3: Channel delay is zero ($R = 0$).
A6.2.4: Operation is unbuffered; that is, a second packet cannot be generated until the previous one has been successfully transmitted.

These assumptions will be relaxed in subsequent models.

Under the above assumptions, a discrete-time queueing network model for unbuffered slotted Aloha is shown in Fig. 6.3. Users at node 1 are idle, and a user at this node can generate a new packet (which implies a transmission attempt) with probability $\sigma$ per slot, with packets assumed to be generated at the beginning of a slot, as before. Users at node 2 have a packet waiting for retransmission, and a user attempts to retransmit its packet with probability $p$ per slot. If exactly one user attempts a transmission from node 1 or a retransmission from node 2 in a slot, this is successful, and the corresponding user jumps to node 1 at the end of the slot. If two or more transmissions and/or retransmissions are attempted in a slot, the corresponding users are involved in a collision, and jump to node 2 at the end of the slot. If no transmissions or retransmissions are attempted in a slot, the users do not jump, but remain at their respective nodes.

The state of this model is

$$x = x^1 \tag{6.30}$$

The state space is

$$\mathbf{X}_N = \{x \in \mathbf{X} | x^1 + x^2 = N\} \tag{6.31}$$

$$\mathbf{X} = \{0, 1, 2, \ldots\} \tag{6.32}$$

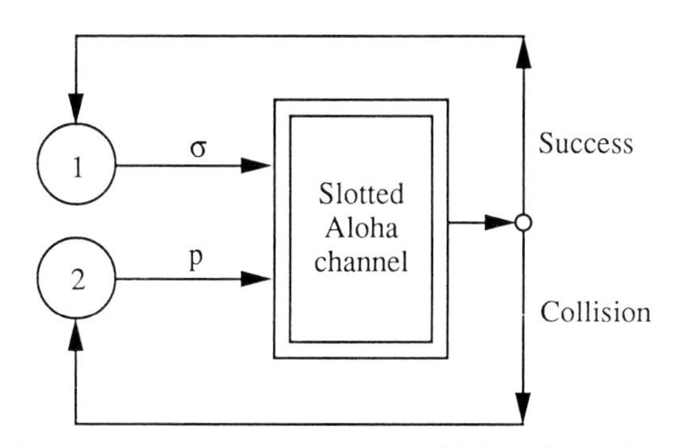


*Figure 6.3 Discrete-time queueing network model for unbuffered slotted Aloha*

The node service rates are

$$\mu_1 = \sigma \tag{6.33}$$

$$\mu_2 = p \tag{6.34}$$

A user that is served at node 1 will be routed back to node 1 if, and only if, non of the other users are served at node 1, and no users are served at node 2. That is

$$r_{11} = (1 - \sigma)^{x^1 - 1}(1 - p)^{x^2} \tag{6.35}$$

A user that is served at node 2 will be routed to node 1 if, and only if, non of the other users are served at node 2, and no users are served at node 1. That is

$$r_{21} = (1 - p)^{x^2 - 1}(1 - \sigma)^{x^1} \tag{6.36}$$

Following directly from this we have

$$r_{12} = 1 - (1 - \sigma)^{x^1 - 1}(1 - p)^{x^2} \tag{6.37}$$

and

$$r_{22} = 1 - (1 - p)^{x^2 - 1}(1 - \sigma)^{x^1} \tag{6.38}$$

Clearly these routing probabilities are state dependent. This means the corresponding flow conservation equations will be non-linear, and so we use EPA to obtain a solution.

Using equations (6.33)–(6.36) in equation (5.40), the equilibrium point equation for node 1 becomes

$$x^1\sigma = x^1\sigma(1 - \sigma)^{x^1 - 1}(1 - p)^{x^2} + x^2 p(1 - p)^{x^2 - 1}(1 - \sigma)^{x^1} \tag{6.39}$$

The constraint equation is

$$x^1 + x^2 = N \tag{6.40}$$

We need not write down the equation for node 2 due to the linear dependence. Using equation (6.40) in (6.39) we get

$$x^1\sigma = x^1\sigma(1 - \sigma)^{x^1 - 1}(1 - p)^{N - x^1} + (N - x^1)p(1 - p)^{N - x^1 - 1}(1 - \sigma)^{x^1} \tag{6.41}$$

Manipulating this, we get a fixed point equation for $x^1$ of the form

$$x^1 = \frac{Np(1 - p)^{N - x^1 - 1}(1 - \sigma)^{x^1}}{\sigma[1 - (1 - \sigma)^{x^1 - 1}(1 - p)^{N - x^1}] + p(1 - p)^{N - x^1 - 1}(1 - \sigma)^{x^1}} \tag{6.42}$$

The conditional throughput, given that the system is at state $x$ is

$$S(x) = x^1\sigma \tag{6.43}$$

Let $x_e^1$ be a solution to equation (6.42). Then the expected value of throughput is

$$E[S(x)] \simeq S(x_e^1) = S = x_e^1 \sigma \tag{6.44}$$

Using Little's result, the mean packet delay can be calculated, as before, to give

$$W = \frac{N}{S} - \frac{1}{\sigma} + 1 \tag{6.45}$$

Before considering numerical results for this model, we shall extend it to account for the case of different customer classes. This is done in the next section.

### 6.2.2 Different customer classes

*Modelling objectives: To show how the performance of a model with statistically different users can be evaluated.*

Here we consider the case where all users have different statistical properties, and so each belongs to a different customer class. That is, assumption A6.2.1 is relaxed. We shall use the conventions and procedure given in Section 5.5.1.

The model for this network is shown in Fig. 6.4. In this model, the same conventions as for the single customer class model apply. A user $i$ at node 1 is served in a slot with probability $\sigma_i$ and jumps from the node just before the end of the slot with this probability. If there are $K$ users in the network, then, under the assumptions made, user $i$ will see the other $K - 1$ users with their equilibrium distribution at the jump. This distribution is approximated by a solution to the EPA equations for the network with the other $K - 1$ users. Thus the probability of at least one transmission/retransmission attempt in a network populated by the other $K - 1$ users will give the probability that user $i$ is involved in a collision,

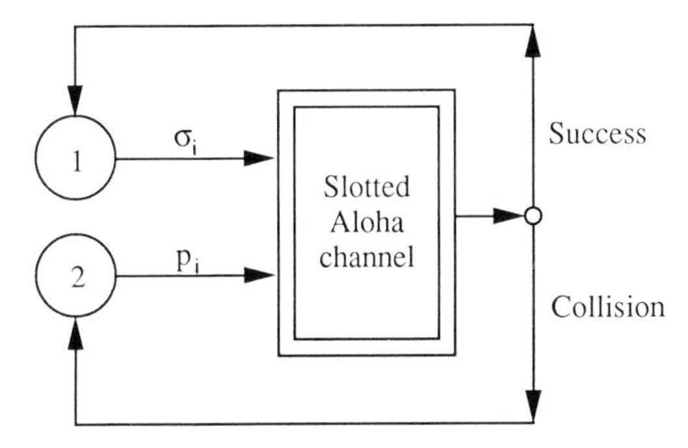


*Figure 6.4 Discrete-time queueing network model for customer class i of slotted Aloha*

and so is routed to node 2. Defining this probability as $\phi_i'(K-1)$, then

$$r_{12} = \phi_i'(K-1) \tag{6.46}$$

$$r_{11} = 1 - \phi_i'(K-1) \tag{6.47}$$

At node 2, a user $i$ is served with probability $p_i$. A similar reasoning to the above gives

$$r_{21} = 1 - \phi_i'(K-1) \tag{6.48}$$

$$r_{22} = \phi_i'(K-1) \tag{6.49}$$

Defining $x_{1,2,\dots,K}(K)$ to be the number of users at node 1 when there are $K$ users in the network, then the equilibrium point equation for user $i$ at node 1 is

$$\begin{aligned} x_{1,2,\dots,K}^{(i)}(K)\sigma_i = {} & x_{1,2,\dots,K}^{(i)}(K)(1-\phi_i'(K-1)) \\ & + (1 - x_{1,2,\dots,K}^{(i)}(K))p_i(1-\phi_i'(K-1)) \end{aligned} \tag{6.50}$$

Note that the constraint equation with only one user in a class implies that user $i$ is at node 2 with probability $1 - x_{1,2,\dots,K}^{(i)}(K)$. This has been incorporated directly into equation (6.50).

Manipulation of equation (6.50) gives

$$x_{1,2,\dots,K}^{(i)}(K) = \frac{p_i(1-\phi_i'(K-1))}{\sigma_i\phi_i'(K-1) + p_i(1-\phi_i'(K-1))} \tag{6.51}$$

Since users are assumed to behave independently, then

$$\phi[x_{1,2,\dots,K}(K)] = 1 - \prod_{i=1}^{K}(1 - (x_{1,2,\dots,K}^{(i)}(K)\sigma_i + (1 - x_{1,2,\dots,K}^{(i)}(K))p_i)) \tag{6.52}$$

In equation (6.51), $\phi_i'(K-1)$ can be obtained from equation (6.52) using equation (5.81).

Equations (6.51) and (6.52) can thus be solved for increasing population levels using the procedure given in Section 5.5.1. The throughput of user $i$ is then

$$S(x_{1,2,\dots,N}^{(i)}(N)) = x_{1,2,\dots,N}^{(i)}(N)\sigma_i \qquad i = 1, 2, \dots, N \tag{6.53}$$

and the total expected throughput of the network is

$$S(x_{1,2,\dots,N}(N)) = \sum_{i=1}^{N} S(x_{1,2,\dots,N}^{(i)}(N)) \tag{6.54}$$

Mean packet delay for user $i$, $W(x_{1,2,\dots,N}^{(i)}(N))$, is obtained using Little's result as

$$W(x_{1,2,\dots,N}^{(i)}(N)) = \frac{1}{S(x_{1,2,\dots,N}^{(i)}(N))} - \frac{1}{\sigma_i} + 1 \qquad i = 1, 2, \dots, N \tag{6.55}$$

Mean packet delay for the network is thus

$$W_{1,2,\ldots,N}(N) = \frac{1}{N} \sum_{i=1}^{N} W(x_{1,2,\ldots,N}^{(i)}(N)) \tag{6.56}$$

The performance of this recursive form of EPA (REPA) is shown compared with the standard EPA method, and also with simulations, in Figs. 6.5 and 6.6, for a network having 16 statistically identical users, each with $\sigma_i = \sigma = 0.01$. Figures 6.5 and 6.6 show how throughput and mean packet delay respectively vary with retransmission probability $p_i = p$. For the standard EPA results, equations (6.42), (6.44) and (6.45) were used. For REPA, the relevant equations are (6.51), (6.54) and (6.56).

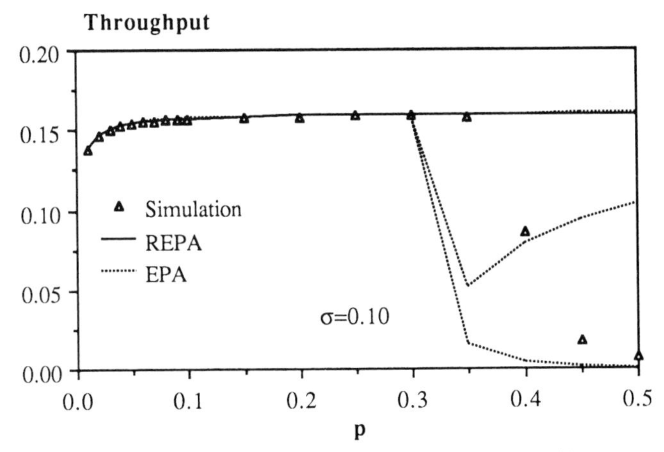


*Figure 6.5 Throughput for the slotted Aloha model – bistable case*

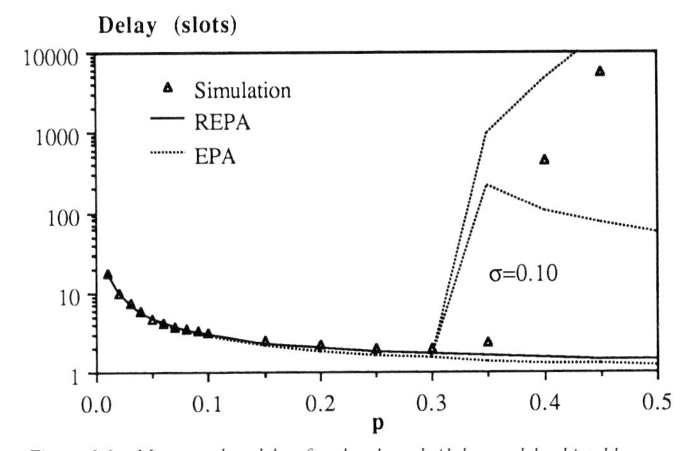


*Figure 6.6 Mean packet delay for the slotted Aloha model – bistable case*

The case chosen is of particular interest in that the network becomes unstable (locally bistable) just beyond $p = 0.35$. (The instability problem for slotted Aloha will not be discussed here, since an excellent treatment of this can be found in many textbooks; for example [TANE 81], [SCHW 87], [TASA 86].) With the network locally bistable, the EPA equations have three solutions, and the trajectories of all three are shown in the figures. It can be seen that both EPA and REPA give an excellent match to simulations for both throughput and delay up to the point where the network becomes unstable. The REPA results then tend to follow the trajectory of the locally stable equilibrium point associated with the higher value of the throughput (or lower value of delay).

With $\sigma$ increased to 0.1, throughput and mean packet delay are again shown in Figs 6.7 and 6.8 respectively. The system is now stable and both EPA and REPA give an accurate and almost indistinguishable match to the simulations.

The performance of the REPA algorithm in the case of 16 users, each belonging to a different customer class, is shown in Table 6.1. Here the comparison is with simulations only, since the standard form of EPA cannot be applied in this situation. Table 6.1(a) shows the network in an underloaded condition, Table 6.1(b) with an intermediate load, and Table 6.1(c) when the load is approximately optimum (near maximum throughput). These results are typical of the accuracy that can be achieved when a network is operating under stable conditions. It can be seen that REPA gives excellent predictions of performance, both for the individual users and the overall network, with the accuracy deteriorating slightly as the load is increased.

Further details of the REPA method can be found in [WOOD 93A] and [WOOD 93B].

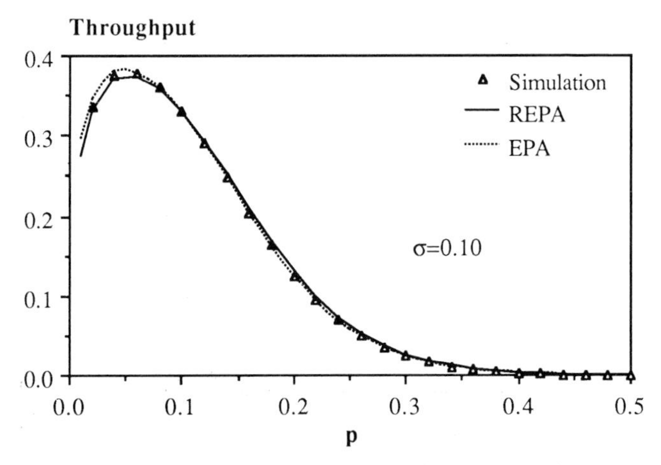


*Figure 6.7 Throughput for the slotted Aloha model – stable case*

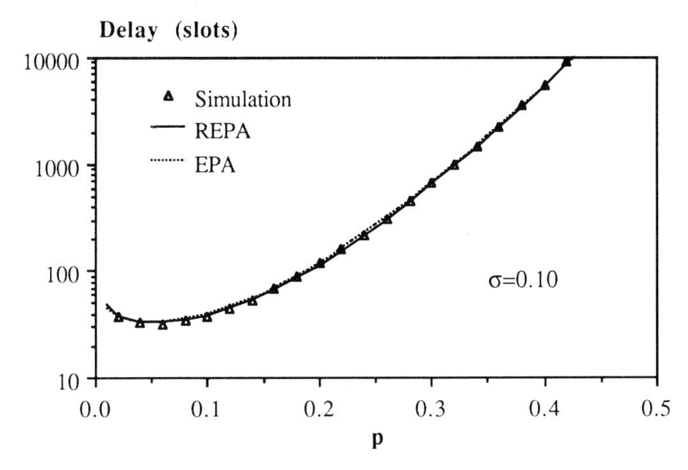


*Figure 6.8 Mean packet delay for the slotted Aloha model – stable case*

### 6.2.3 Finite channel delay

*Modelling objectives: To show how EPA can be applied to models with a multidimensional state space, and to introduce a stochastic transformation that can be used to simplify such models.*

In this section we revert to the single customer class model, but relax assumptions A6.2.2 and A6.2.3. That is, messages can now consist of multiple packets, and there is a finite, non-zero delay on the channel.

The corresponding queueing network model is shown in Fig. 6.9. Before analysing this model however, a simplification will be effected to that part of the model that comprises the channel input; that is, nodes 0 and $2R + 1$. We shall carry out a stochastic transformation on this part of the model so that the channel has a single rather than a dual input. This is shown diagrammatically in Fig. 6.10. Nodes 0 and $2R + 1$, which form a parallel input to the channel in Fig. 6.10(a) are replaced by the tandem arrangement consisting of nodes $0'$ and $2R + 1'$ shown in Fig. 6.10(b), such that *the stochastic behaviour of the two models is identical.*

The problem is how to choose the routing probability $x$ in Fig. 6.10(b) so that the behaviour of the node arrangements in Figs. 6.10(a) and 6.10(b) is stochastically equivalent. The solution to this problem is a good illustration of yet another use of generating functions.

Let $U_1$, $W_1$ and $W_2$ be random variables representing the number of slots a user remains at nodes $2R + 1$,, 0, and $0'$, respectively. Also, let $U_2$ be a random variable representing the number of slots from the instant a user enters either node $2R + 1'$ or $0'$, until that user leaves node $0'$. This is shown diagrammatically in Fig. 6.11. Also, let the generating functions

**Table 6.1(a)** PERFORMANCE RESULTS FOR SLOTTED ALOHA WITH DIFFERENT CUSTOMER CLASSES (UNDERLOADED)

| | | | Delay | | | Throughput | | |
|---|---|---|---|---|---|---|---|---|
| *User i* | $\sigma_i$ | $p_i$ | *Simul* | *REPA* | *% err* | *Simul* | *REPA* | *% err* |
| 1 | 0.0070 | 0.0010 | 171.918 | 164.600 | −4.26% | 0.0033 | 0.0033 | −0.74% |
| 2 | 0.0074 | 0.0022 | 78.604 | 74.460 | −5.27% | 0.0047 | 0.0048 | 2.52% |
| 3 | 0.0078 | 0.0034 | 47.701 | 48.180 | 1.00% | 0.0057 | 0.0057 | −0.08% |
| 4 | 0.0082 | 0.0046 | 36.191 | 35.680 | −1.41% | 0.0064 | 0.0064 | 0.17% |
| 5 | 0.0086 | 0.0058 | 28.408 | 28.370 | −0.13% | 0.0070 | 0.0070 | −0.13% |
| 6 | 0.0090 | 0.0070 | 23.095 | 23.590 | 2.14% | 0.0075 | 0.0075 | −0.66% |
| 7 | 0.0094 | 0.0082 | 20.359 | 20.200 | −0.78% | 0.0080 | 0.0080 | −0.29% |
| 8 | 0.0098 | 0.0094 | 17.800 | 17.690 | −0.62% | 0.0084 | 0.0084 | 0.10% |
| 9 | 0.0102 | 0.0106 | 15.914 | 15.750 | −1.03% | 0.0088 | 0.0089 | 0.47% |
| 10 | 0.0106 | 0.0118 | 14.615 | 14.200 | −2.84% | 0.0093 | 0.0093 | −0.05% |
| 11 | 0.0110 | 0.0130 | 12.941 | 12.940 | −0.01% | 0.0098 | 0.0097 | −0.49% |
| 12 | 0.0114 | 0.0142 | 11.875 | 11.890 | 0.13% | 0.0102 | 0.0101 | −0.42% |
| 13 | 0.0118 | 0.0154 | 10.601 | 11.010 | 3.86% | 0.0106 | 0.0106 | −0.86% |
| 14 | 0.0122 | 0.0166 | 10.122 | 10.253 | 1.29% | 0.0110 | 0.0110 | −0.19% |
| 15 | 0.0126 | 0.0178 | 9.551 | 9.600 | 0.52% | 0.0114 | 0.0114 | 0.06% |
| 16 | 0.0130 | 0.0190 | 9.007 | 9.030 | 0.26% | 0.0117 | 0.0118 | 0.29% |
| Mean | 0.0100 | 0.0100 | 32.419 | 31.720 | −2.16% | 0.1337 | 0.1337 | −0.04% |

Note: The % error is based on the results expressed to 6 decimal places.

**Table 6.1(b)** PERFORMANCE RESULTS FOR SLOTTED ALOHA WITH DIFFERENT CUSTOMER CLASSES (INTERMEDIATE LOAD)

| | | | *Delay* | | | *Throughput* | | |
|---|---|---|---|---|---|---|---|---|
| *User i* | $\sigma_i$ | $p_i$ | *Simul* | *REPA* | *% err* | *Simul* | *REPA* | *% err* |
| 1 | 0.0210 | 0.0010 | 402.233 | 357.700 | −11.07% | 0.0023 | 0.0025 | 8.14% |
| 2 | 0.0222 | 0.0022 | 165.952 | 161.600 | −2.62% | 0.0048 | 0.0049 | 1.60% |
| 3 | 0.0234 | 0.0034 | 103.546 | 104.100 | 0.54% | 0.0069 | 0.0069 | −0.81% |
| 4 | 0.0246 | 0.0046 | 77.937 | 76.640 | −1.66% | 0.0085 | 0.0086 | 0.78% |
| 5 | 0.0258 | 0.0058 | 61.638 | 60.600 | −1.68% | 0.0100 | 0.0102 | 1.20% |
| 6 | 0.0270 | 0.0070 | 49.417 | 50.090 | 1.36% | 0.0117 | 0.0116 | −0.97% |
| 7 | 0.0282 | 0.0082 | 42.575 | 42.680 | 0.25% | 0.0130 | 0.0130 | −0.65% |
| 8 | 0.0294 | 0.0094 | 36.917 | 37.170 | 0.69% | 0.0143 | 0.0143 | −0.49% |
| 9 | 0.0306 | 0.0106 | 32.789 | 32.910 | 0.37% | 0.0155 | 0.0155 | −0.35% |
| 10 | 0.0318 | 0.0118 | 29.524 | 29.520 | −0.01% | 0.0167 | 0.0167 | −0.30% |
| 11 | 0.0330 | 0.0130 | 26.159 | 26.770 | 2.33% | 0.0181 | 0.0178 | −1.18% |
| 12 | 0.0342 | 0.0142 | 24.144 | 24.480 | 1.39% | 0.0191 | 0.0190 | −0.69% |
| 13 | 0.0354 | 0.0154 | 21.666 | 22.550 | 4.08% | 0.0205 | 0.0201 | −1.90% |
| 14 | 0.0366 | 0.0166 | 20.453 | 20.900 | 2.19% | 0.0213 | 0.0212 | −0.79% |
| 15 | 0.0378 | 0.0178 | 18.744 | 19.480 | 3.93% | 0.0226 | 0.0223 | −1.42% |
| 16 | 0.0390 | 0.0190 | 17.704 | 18.230 | 2.97% | 0.0235 | 0.0233 | −0.89% |
| Mean | 0.0300 | 0.0100 | 70.712 | 67.830 | −4.08% | 0.2290 | 0.2276 | −0.61% |

Note: The % error is based on the results expressed to 6 decimal places.

**Table 6.1(c)** PERFORMANCE RESULTS FOR SLOTTED ALOHA WITH DIFFERENT CUSTOMER CLASSES (NEAR OPTIMUM LOAD)

| | | | Delay | | | Throughput | | |
|---|---|---|---|---|---|---|---|---|
| *User i* | $\sigma_i$ | $p_i$ | *Simul* | *REPA* | *% err* | *Simul* | *REPA* | *% err* |
| 1 | 0.0210 | 0.0100 | 102.164 | 97.560 | −4.51% | 0.0068 | 0.0069 | 2.75% |
| 2 | 0.0222 | 0.0220 | 46.501 | 43.750 | −5.92% | 0.0111 | 0.0114 | 2.97% |
| 3 | 0.0234 | 0.0340 | 30.484 | 28.210 | −7.46% | 0.0138 | 0.0143 | 3.28% |
| 4 | 0.0246 | 0.0460 | 22.753 | 20.870 | −8.27% | 0.0160 | 0.0165 | 3.03% |
| 5 | 0.0258 | 0.0580 | 18.039 | 16.600 | −7.98% | 0.0180 | 0.0184 | 2.19% |
| 6 | 0.0270 | 0.0700 | 15.242 | 13.810 | −9.40% | 0.0195 | 0.0201 | 2.78% |
| 7 | 0.0282 | 0.0820 | 12.978 | 11.840 | −8.77% | 0.0211 | 0.0216 | 2.56% |
| 8 | 0.0294 | 0.0940 | 11.614 | 10.385 | −10.58% | 0.0223 | 0.0230 | 3.44% |
| 9 | 0.0306 | 0.1060 | 10.350 | 9.261 | −10.52% | 0.0238 | 0.0244 | 2.52% |
| 10 | 0.0318 | 0.1180 | 9.386 | 8.368 | −10.84% | 0.0251 | 0.0258 | 2.74% |
| 11 | 0.0330 | 0.1300 | 8.630 | 7.642 | −11.44% | 0.0264 | 0.0271 | 2.60% |
| 12 | 0.0342 | 0.1420 | 7.941 | 7.040 | −11.34% | 0.0275 | 0.0283 | 2.93% |
| 13 | 0.0354 | 0.1540 | 7.418 | 6.532 | −11.94% | 0.0288 | 0.0296 | 2.80% |
| 14 | 0.0366 | 0.1660 | 6.938 | 6.099 | −12.09% | 0.0299 | 0.0308 | 3.12% |
| 15 | 0.0378 | 0.1780 | 6.572 | 5.725 | −12.89% | 0.0313 | 0.0321 | 2.44% |
| 16 | 0.0390 | 0.1900 | 6.127 | 5.398 | −11.89% | 0.0325 | 0.0333 | 2.28% |
| Mean | 0.0300 | 0.1000 | 20.196 | 18.690 | −7.46% | 0.3539 | 0.3637 | 2.76% |

Note: The % error is based on the results expressed to 6 decimal places.

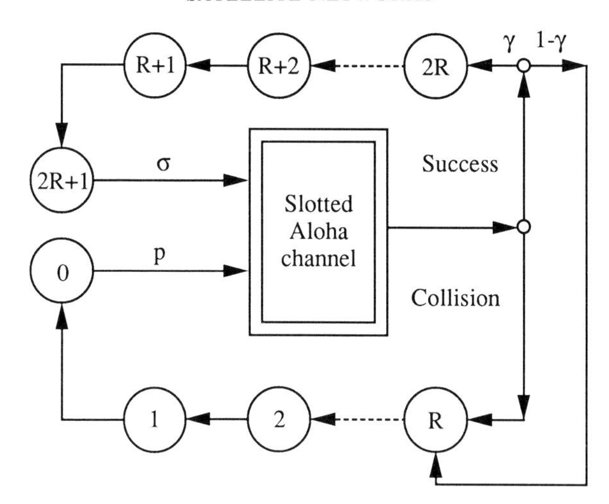


*Figure 6.9 Discrete-time queueing network model for slotted Aloha with multipacket messages and finite channel delay*

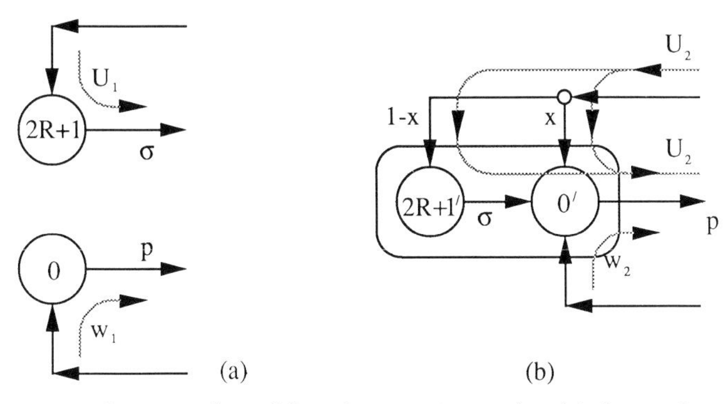


*Figure 6.10 Stochastic transform of channel input; (a) original model, (b) transformed model*

of $U_1$, $U_2$, $W_1$ and $W_2$ be denoted by $G_{U_1}(z)$, $G_{U_2}(z)$, $G_{W_1}(z)$ and $G_{W_2}(z)$, respectively. Then from Fig. 6.10 we have

$$G_{U_1}(z) = \frac{\sigma z}{1-(1-\sigma)z} \tag{6.57}$$

$$G_{W_1}(z) = \frac{pz}{1-(1-\sigma)z} \tag{6.58}$$

$$G_{W_2}(z) = \frac{pz}{1-(1-\sigma)z} \tag{6.59}$$

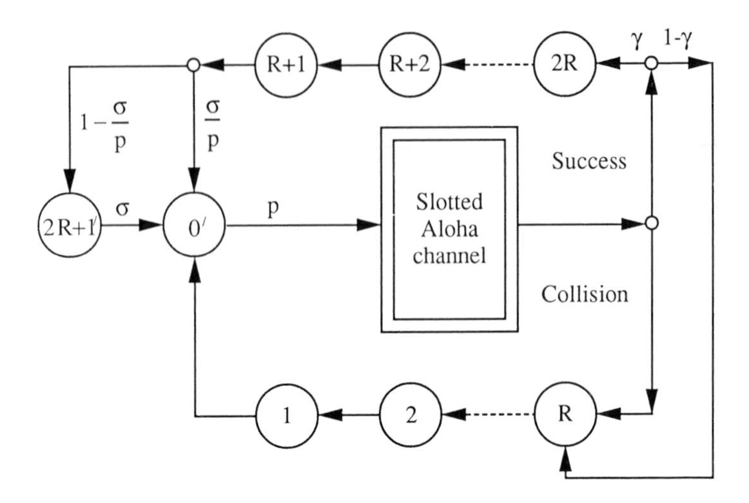


*Figure 6.11 Modified model for slotted Aloha*

That is to say $U_1$, $W_1$ and $W_2$ are all geometrically distributed random variables.

Now using the properties of generating functions, and equating $G_{U_2}(z)$ to $G_{U_1}(z)$, we have

$$(1-x)\frac{\sigma z}{1-(1-\sigma)z}G_{W_2}(z) + xG_{W_2}(z) = G_{U_1}(z) \tag{6.60}$$

The value of $x$ that solves equation (6.60) must therefore make the generating functions of the random variables $U_1$ and $U_2$ the same, hence they must have the same probability distribution. Substitution of $G_{U_1}(z)$ and $G_{W_2}(z)$ in equation (6.60) gives

$$x = \frac{\sigma}{p} \tag{6.61}$$

The complete transformed model is shown in Fig. 6.11. The channel now has a single input from node $0'$. This greatly simplifies the application of EPA without otherwise impairing the stochastic behaviour of the system.

We now have a situation such that a user that is served at node $0'$ will successfully transmit a packet if, and only if, no other users are served at node $0'$ in the same slot. If this is the last packet of a message, the user is routed back to either node $0'$ or $2R + 1'$ via a channel delay of $R$ slots which is modelled by the nodes $2R$ $2R - 1, \ldots, R + 1$, each of which has a service rate of 1. If there are more packets in the message, then the user is routed to node R, to return to node $0'$ after an $R$ slot delay as before.

If two or more users are served at node $0'$ in the same slot, these are involved in a collision and routed to node $R$. After an $R$ slot delay they return to node $0'$ for their packets to be retransmitted.

The state vector of this model is

$$x = (x^{0'}, x^1, x^2, \ldots, x^{2R}) \tag{6.62}$$

The state space is

$$\mathbf{X}_N = \left\{ x \in \mathbf{X} | x^{0'} + x^{2R+1'} + \sum_{i=1}^{2R} x^i = N \right\} \tag{6.63}$$

$$\mathbf{X} = \{0, 1, 2, \ldots\}^{2R+1} \tag{6.64}$$

The EPA equations are

$$x^{2R+1'}\sigma = x^{R+1}\left(1 - \frac{\sigma}{p}\right) \tag{6.65}$$

$$x^{R+1} = x^{R+2} = \cdots = x^{2R} = x^{0'}p(1-p)^{x^{0'-1}-1}\gamma \tag{6.66}$$

$$x^1 = x^2 = \cdots = x^R = x^{0'}p(1-(1-p)^{x^{0'}-1}) + x^{0'}p(1-p)^{x^{0'}-1}(1-\gamma) \tag{6.67}$$

The constraint equation is

$$x^{0'} + x^{2R+1'} + \sum_{i=1}^{2R} x^i = N \tag{6.68}$$

The equation for node $0'$ has not been used since it is linearly dependent on the others.

Equation (6.68) can be used to eliminate $x^{2R+1'}$ from equation (6.65), after which equations (6.65), (6.66) and (6.67) can be solved in terms of $x^{0'}$ to give

$$x^{0'} = \frac{N\dfrac{\sigma}{p}}{\dfrac{\sigma}{p}(1+Rp) + \gamma\left(1-\dfrac{\sigma}{p}\right)(1-p)^{x^{0'}-1}} \tag{6.69}$$

Throughput is easily obtained as

$$S(x) = x^{0'}p(1-p)^{x^{0'}-1} \tag{6.70}$$

As before, if $x_e^{0'}$ is a solution to equation (6.69), then the expected value of throughput is obtained as

$$E[S(x)] \simeq S(x_e^{0'}) = S = x_e^{0'}p(1-p)^{x_e^{0'}-1} \tag{6.71}$$

Mean message delay in slots is given by Little's result as

$$W = \frac{N}{\gamma S} - \frac{1}{\sigma} + 1 \tag{6.72}$$

Note that by setting $R = 0$, $\gamma = 1$ in equation (6.69) this model will given an identical result to the model with zero channel delay and single packet messages considered in Section 6.2.1. That equation (6.69) with $R = 0$, $\gamma = 1$ does not reduce to equation (6.42) is a consequence of the fact that we are here using a transformed model, and the random variables $x^{0'}$ in the present model, and $x^1$ in that of Section 6.2.1 are not the same thing. Note also that equation (6.72) for mean message delay does not explicitly contain $R$, the channel propagation delay. It is, of course, contained implicitly in $S$, since this latter is a function of $x^{0'}$.

Figures 6.12 and 6.13 show typical performance accuracy of this model

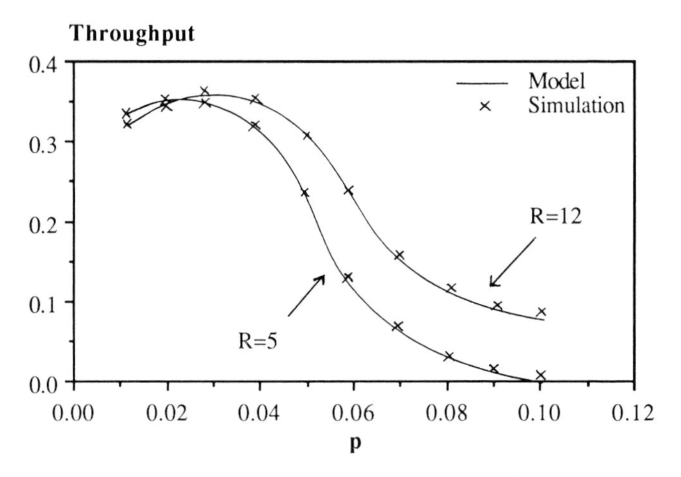


*Figure 6.12 Throughput for the slotted Aloha model with finite channel delay*

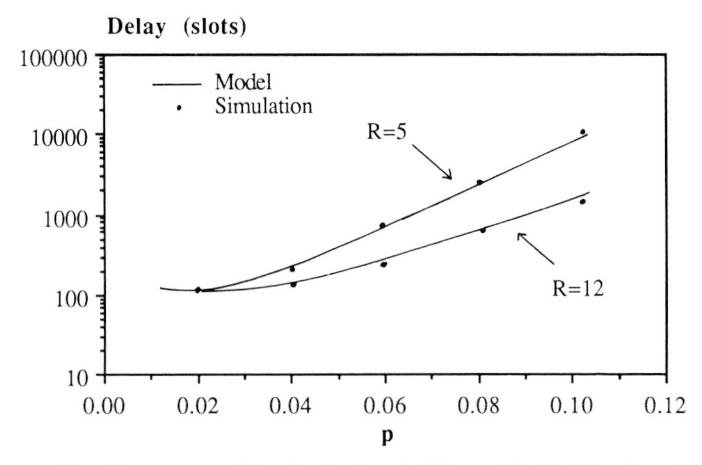


*Figure 6.13 Mean message delay for the slotted Aloha model with finite channel delay*

for a network with different channel delays of $R = 5$ and $R = 12$. The other parameter values are $N = 100$, $\sigma = 0.006$, and $\gamma = 1$. The results have been compared with a simulation taken over $10^5$ slots. Again it can be seen that accuracy is good provided the protocol is stable.

A similar analysis to the one given here can be found in [TASA 86].

## 6.3 CODE DIVISION MULTIPLE ACCESS

*Modelling objectives: To show how the statistics of spread spectrum modulation schemes can be included in the discrete-time queueing models.*

In slotted Aloha, when two or more users attempt a packet transmission in the same slot, the mutual interference between the packets causes their information content to be destroyed, and a retransmission must be initiated. If each user is assigned a unique code to be modulated by its data, such that the set of $N$ codes are (approximately) mutually orthogonal (each code has low cross-correlation with all the other codes), then two or more users may transmit in the same slot without interference. Such schemes are known as *code division multiple access* (CDMA), and the throughputs that can be achieved will clearly be greater than 'binary' type schemes (collision, or no collision) such as slotted Aloha.

In effect, CDMA is an application of spread spectrum modulation, and can be implemented in a number of different ways (direct sequence spread; frequency hopping; time hopping). Details of these schemes can be found in many good textbooks [PROA 89], [HAYK 88], [SKLA 88], [LEE 88], [COOP 86], and will not be considered here. What we shall consider is how to incorporate their effect into a discrete-time queueing network model for a satellite packet broadcast channel.

The model we shall examine is shown in Fig. 6.14, and is an extension of the modified slotted Aloha model of Section 6.2.3. The operation of the model of Fig. 6.14 is identical with that of Fig. 6.11, with one exception; this being the inclusion of the 'fixed assigned' performance parameters of the CDMA scheme, which are assumed to be known. That is, the error probabilities for a given number, $m$, of simultaneous transmissions in the same slot are available. In this respect, $P_C(m)$ in Fig. 6.14 represents the probability of a correct packet reception with $m$ simultaneous transmissions in the same slot.

The other change in Fig. 6.14 is that nodes $0'$ and $2R + 1'$ in Fig. 6.11 are relabelled 0 and $2R + 1$ for convenience.

Then with reference to Fig. 6.14, we see that if $m$ users are served at node 0 in a slot, then each of these is successfully transmitted with probability $P_c(m)$, and is routed accordingly.

Let $M$ be a random variable representing the number of attempted transmissions in a slot, and $K$ be a random variable representing the number of successfully received packets, both given that the system is at

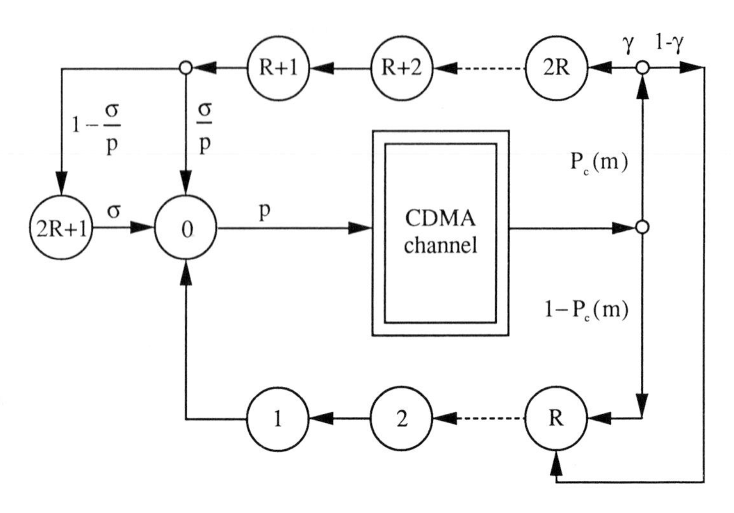


*Figure 6.14 Discrete-time queueing network model for random access CDMA*

state $x$. Then the conditional distribution of $K$ is

$$P(K = k|M = m) = \binom{m}{k} P_c^k(m) P_E^{m-k}(m) \tag{6.73}$$

where $P_E(m) = 1 - P_c(m)$.

The conditional mean value of throughput, given that the system is at state $x$, is now given by

$$\begin{aligned} S(x) &= E[K] \\ &= E[E(K|M]]; \qquad M \leq x^0 \\ &= E\left[\sum_{k=0}^{M} k \binom{M}{k} P_c^k(M) P_E^{M-k}(M)\right] \\ &= E[MP_c(M)] \\ &= \sum_{m=1}^{x^0} m \binom{x^0}{m} p^m (1-p)^{x^0-m} P_c(m) \end{aligned} \tag{6.74}$$

This can be approximated by a solution to the EPA equations as before.

The state vector of the model is

$$x = (x^0, x^1, \ldots, x^{2R}) \tag{6.75}$$

The state space is

$$\mathbf{X}_N = \left\{ x \in \mathbf{X} \,\middle|\, \sum_{i=0}^{2R+1} x^i = N \right\} \tag{6.76}$$

$$\mathbf{X} = \{0, 1, 2, \ldots\}^{2R+1} \tag{6.77}$$

The EPA equations are

$$x^{2R+1}\sigma = x^{R+1}\left(1 - \frac{\sigma}{p}\right) \tag{6.78}$$

$$x^{R+1} = x^{R+2} = \cdots = x^{2R} = \gamma S(x) \tag{6.79}$$

$$x^1 = x^2 = \cdots = x^R = (1 - \gamma)S(x) + x^0 p - S(x) \tag{6.80}$$

The constraint equation is

$$\sum_{i=0}^{2R+1} x^i = N \tag{6.81}$$

From equations (6.78) to (6.81) we obtain

$$x^0 = \frac{N - \left(1 - \frac{\sigma}{p}\right)\frac{\gamma}{\sigma}S(x)}{1 + Rp} \tag{6.82}$$

Since $S(x)$ is a function of $x^0$ (see equation (6.74)), then we have a fixed point equation that can be solved to obtain an equilibrium value of $x^0$, say $x_e^0$. The expected value of throughput can then be approximated as

$$E[S(x)] \simeq S(x_e^0) = S = \sum_{m=1}^{x_e^0} m\binom{x_e^0}{m}p^m(1 - p)^{x_e^0 - m}P_c(m) \tag{6.83}$$

Note that $x_e^0$ in equation (6.83) must be taken as the nearest integer value.

As with the slotted Aloha model of Section 6.2.3, the mean message delay is given by

$$W = \frac{N}{S\gamma} - \frac{1}{\sigma} + 1 \tag{6.84}$$

It is apparent that this model reduces to the previous slotted Aloha model if we note that for slotted Aloha

$$P_c(m) = \begin{cases} 1, & m = 1 \\ 0, & otherwise \end{cases} \tag{6.85}$$

In other words, for slotted Aloha, $P_c(m)$ acts as a switching function in that if only one transmission is attempted in a slot the corresponding user is successful and routed to node $R$ or $2R$ depending on whether the user's message has more packets or not, respectively. If more than one user attempts a transmission, these are routed directly to node $R$. If no users attempt a transmission then the value of $P_c(m)$ is immaterial.

Using equation (6.85) in equation (6.83) we see that

$$S = x_e^0 p(1 - p)^{x_e^0 - 1} \tag{6.86}$$

This is exactly the slotted Aloha result of equation (6.71).

CDMA can also be used on a *fixed assignment* basis in which a fixed number of users, $M_0$, transmit packets in *every* time slot. In this case we have

$$E[K|M] = \begin{cases} M_0 P_c(M_0), & M = M_0 \\ 0, & \textit{otherwise} \end{cases} \tag{6.87}$$

Since $P(M = M_0) = 1$ then

$$S = E[K] = M_0 P_c(M_0) \tag{6.88}$$

To assess the performance of the CDMA model against simulations, we require to consider some specific modulation scheme in order to evaluate $P_c(M)$. As an example we shall use *direct sequence spread binary phase shift keying* (DS/BPSK) as the spread spectrum modulation scheme [HAYK 88]. For this, with $m$ users attempting transmission in the same slot the *average probability of a bit error*, $P_e$, is given approximately by

$$P_e \simeq \frac{1}{2} erfc\left[\sqrt{\frac{G_p}{m-1}}\right] \tag{6.89}$$

Here, $G_p$ is the *processing gain* which will be a known quantity for a given scheme, and erfc(.) is the complementary error function, defined in the usual way as

$$erfc(u) = \frac{2}{\sqrt{\pi}} \int_u^\infty \exp(-z^2)\, dz \tag{6.90}$$

A derivation of equation (6.89) can be found in [PROA 89].

With packets of length $L$ bits, the *average probability of packet error* with $m$ simultaneous users on the channel, and assuming the bits are independent, is given by a binomial expansion as

$$P_E(m) = \binom{L}{1} P_e(1 - P_e)^{L-1} + \binom{L}{2} P_e^2(1 - P_e)^{L-2} + \cdots \tag{6.91}$$

Typically, $P_e$ is small (e.g. $10^{-6}$), and we can use the approximation

$$\begin{aligned} P_E(m) &\simeq L P_e \\ &= \frac{L}{2} erfc\left[\sqrt{\frac{G_p}{m-1}}\right] \end{aligned} \tag{6.92}$$

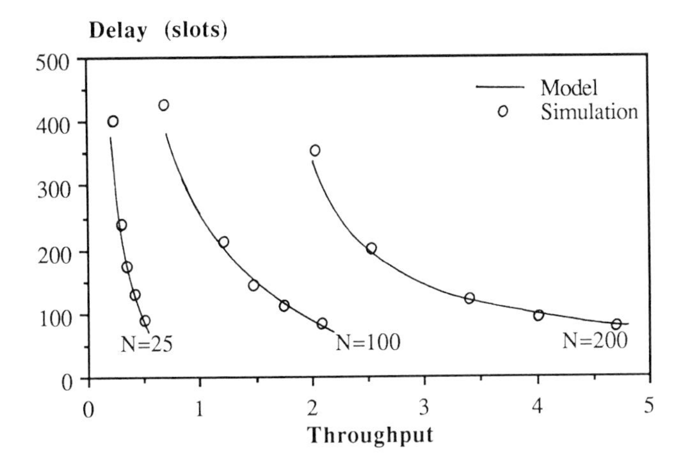


*Figure 6.15 Delay-throughput characteristics for the CDMA model*

Then the probability of *correct* packet reception, with $m$ simultaneous users on the channel, becomes

$$P_c(m) \simeq 1 - \frac{L}{2} erfc\left[\sqrt{\frac{G_p}{m-1}}\right] \tag{6.93}$$

Typically a DS/BPSK spread spectrum modulation scheme would use a processing gain of at least 1000.

Since erfc(.) is a monotonic decreasing function of its argument, as $m$ increases then the probability of correct reception, $P_c(m)$, decreases. This is due to the imperfect cross correlation properties of the codes assigned to the users. Thus the more users that pile up in a slot, the greater the chance of their packets being received incorrectly. In other words, the users' packets act like additive noise on the channel.

Delay-throughput characteristics in which the model is compared with a $10^5$ slot simulation are shown in Fig. 6.15. The parameters used in this are $N = 25$, 100, 200, $\sigma = 0.01$, $\gamma = 0.25$, $p = 0.01$–$0.15$, $L = 1125$, $G_p = 1000$, and a channel propagation delay of 0.27 s (which gives $R = 12$).

The model is seen to match well with the simulations. It is interesting to note that throughputs approaching 5 can be achieved in the 200 user system. This is well over ten times greater than the equivalent slotted Aloha system, which would produce a maximum throughput of around the infinite user limit of $1/e$ for this 200 user system.

An exact Markov analysis for random access CDMA for the case of zero channel delay and single packet messages can be found in [RAYC 81].

## 6.4 BUFFERED SLOTTED ALOHA

*Modelling objectives: To introduce some simplifications that can be used to model systems with both buffered users and finite channel delays; that is, complicated systems with a multidimensional state space.*

In this section we consider again the slotted Aloha protocol, but this time assume that each user has a finite length buffer for storing packets that are awaiting transmission/retransmission. Assumptions A6.1 to A6.8 apply, in addition to the following:

A6.4.1: There is a single user class.
A6.4.2: Messages consist of single packets ($\gamma = 1$).
A6.4.3: Each user maintains a buffer of maximum capacity $J$ packets.
A6.4.4: There is a delayed first transmission; that is, any new packets that are generated are placed in the buffer for one slot before any transmission is attempted. This transmission is then attempted with probability $p$.
A6.4.5: Each user is allowed to generate only one new packet during that user's transmission period; this packet generation is assumed to occur at the start of the last slot of a transmission period with probability $(R + 1)\sigma$.

These last two assumptions are made to simplify the modelling. The performance of a protocol under the delayed first transmission assumption A6.4.4 has been shown to be little different from that of the same protocol where the first transmission is not delayed [TASA 86]. Assumption A6.4.5, that a user is restricted to generate, at most, one new packet during a packet transmission period perhaps needs a little more justification. If $Y$ is a random variable representing the number of packets generated by a user during a transmission period (this will be $R + 1$ slots in length, consisting of 1 slot actual transmission time, plus $R$ slots propagation time), then $Y$ will have a binomial distribution with parameters $(R + 1, \sigma)$. Assumption A6.4.5 means that we are approximating this distribution with a single Bernoulli trial having probability $(R + 1)\sigma$ of success. Provided that $\sigma$ is small, then the error will not be significant. Furthermore, the error will tend to zero as $R \rightarrow 0$, so that for zero propagation delay the approximation is exact.

Under the above assumptions, a discrete-time queueing network model for slotted Aloha with buffered users is shown in Fig. 6.16.

Users at node $B_0$ are idle and have no packets in their buffer. They can generate new packets with probability $\sigma$, in which case they jump to node $B_1$ at the end of the slot in which a packet is generated. Users at nodes $B_i$, $1 \leq i \leq J$, have $i$ packets in their buffer, and they attempt transmission of one of these packets with probability $p$ in each slot. A user at node $B_i$ that successfully transmits a packet in a slot jumps to node $T_{i1}$, $1 \leq i \leq J$,

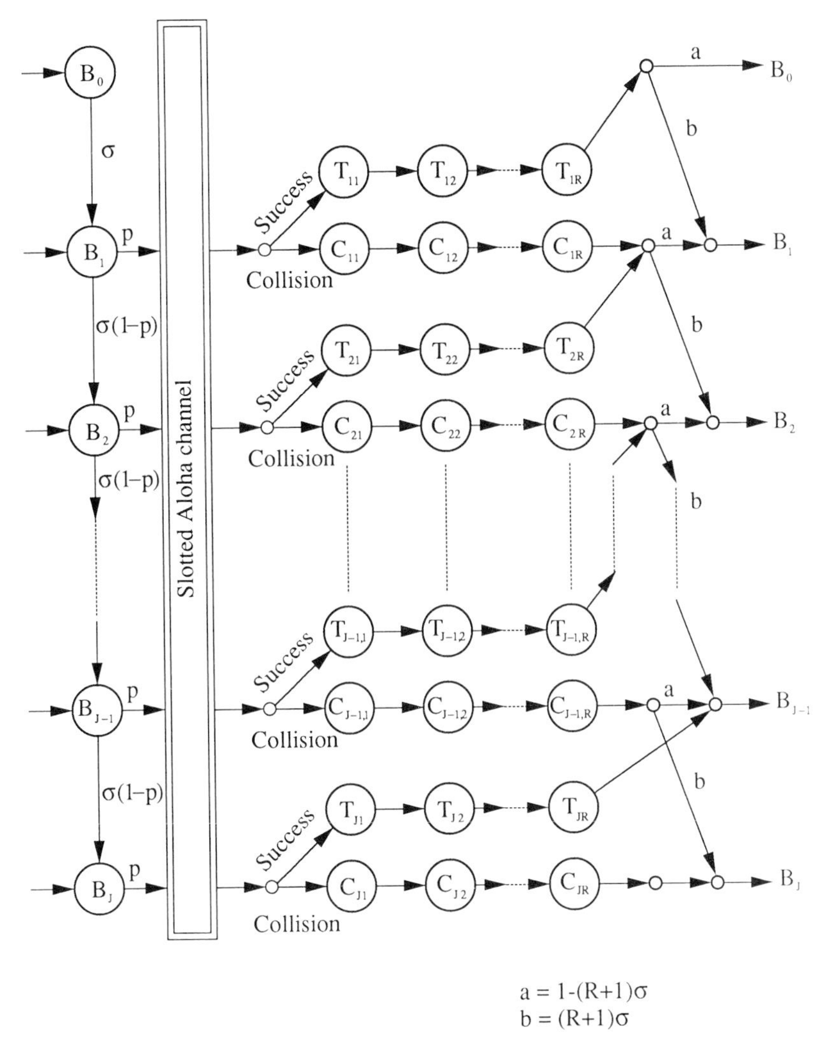


*Figure 6.16 Discrete-time queueing network model for slotted Aloha with buffered users and finite channel delay*

at the end of the slot. A user at node $B_i$ whose packet is involved in a collision jumps to node $C_{i1}$, $1 \leq i \leq J$, at the end of the slot.

Users at node $T_{i1}$, $2 \leq i \leq J - 1$, after an $R$ slot channel delay represented by node $T_{i1}, T_{i2}, \ldots, T_{iR}$, jump to node $B_{i-1}$ if a new packet has not been generated during the transmission period (probability $1 - (R + 1)\sigma$, or jump to node $B_i$ if a new packet has been generated during the transmission period (probability $(R + 1)\sigma$). Users at node $T_{J1}$, after an $R$ slot channel delay represented by nodes $T_{J1}, T_{J2}, \ldots, T_{JR}$, will always jump to node $B_{J-1}$. This is due to assumption A6.4.5, so that if a new packet has been generated during the transmission period, this packet will be lost due to the buffer being full.

Users at node $C_{i1}$, $1 \leq i \leq J-1$, have a collided packet on the channel, and after an $R$ slot delay represented by nodes $C_{i1}, C_{i2}, \ldots, C_{iR}$, jump to node $B_i$ if no new packet has been generated during the transmission period (probability $1-(R+1)\sigma$), or jump to node $B_{i+1}$ if a new packet has been generated (probability $(R+1)\sigma$). Users at node $C_{J1}$, after an $R$ slot delay represented by nodes $C_{J1}, C_{J2}, \ldots, C_{JR}$, always jump to node $B_J$ due to the full buffer.

A user at node $B_i$, $1 \leq i \leq J-1$, that does not attempt transmission in a slot (probability $1-p$) will remain at node $B_i$ if no new packet is generated (probability $1-\sigma$), or jump to node $B_{i+1}$ if a new packet is generated (probability $\sigma$). A user at node $B_J$ must always remain at node $B_J$ if no packet transmission is attempted in a slot, due to the full buffer.

A user at node $B_i$, $1 \leq i \leq J$, that attempts to transmit a packet in a slot will be successful provided that none of the other users at nodes $B_i$, $1 \leq i \leq J$, also attempt a transmission in the same slot. Let $b_i$, $0 \leq i \leq J$, $t_{ij}$, $1 \leq i \leq J, 1 \leq j \leq R$, and $c_{ij}$, $1 \leq i \leq J, 1 \leq j \leq R$, be random variables representing the number of users at nodes $B_i$, $T_{ij}$ and $C_{ij}$, respectively. Then the state vector for this model is

$$x = (b_i, t_{ij}, c_{ij}; 1 \leq i \leq J, 1 \leq j \leq R) \tag{6.94}$$

The state space is

$$\mathbf{X}_N = \left\{ x \in \mathbf{X} | b_0 + \sum_{i=1}^{J} \left[ b_i + \sum_{j=1}^{R} (t_{ij} + c_{ij}) \right] = N; \right.$$

$$\left. t_{ij} \in \{0, 1\};\ \textit{if } t_{ij} = 1, \textit{ then } c_{ij} = 0;\ \textit{if } c_{ij} \neq 0, \textit{ then } t_{ij} = 0 \right\} \tag{6.95}$$

$$\mathbf{X} = \{0, 1, 2, \ldots\}^{J(2R+1)} \tag{6.96}$$

Now define

$$M = \sum_{i=1}^{J} b_i \tag{6.97}$$

and let $S_i(x)$ be the conditional probability that a user at node $B_i$, $1 \leq i \leq J$, successfully transmits a packet, given that the system is at state $x$. Then

$$S_i(x) = b_i p(1-p)^{M-1} \tag{6.98}$$

Thus the conditional throughput at state $x$ is

$$S(x) = \sum_{i=1}^{J} S_i(x) = Mp(1-p)^{M-1} \tag{6.99}$$

For convenience, we shall write $S_i(x) = S_i$. Then the equilibrium point equations are

$$b_0\sigma = t_{1R}(1 - (R + 1)\sigma) \tag{6.100}$$

$$t_{i1} = t_{i2} = \cdots = t_{iR} = S_i; \qquad i = 1, 2, \ldots, J \tag{6.101}$$

$$c_{i1} = c_{i2} = \cdots = c_{iR} = b_i p - S_i; \qquad i = 1, 2, \ldots, J \tag{6.102}$$

$$\begin{aligned} b_i(p + (1 - p)\sigma) = {} & (b_{i-1}p - S_{i-1})(R + 1)\sigma + b_{i-1}(1 - p)\sigma + S_i(R + 1)\sigma \\ & + (b_i p - S_i)(1 - (R + 1)\sigma) + S_{i+1}(1 - (R + 1)\sigma) \\ & i = 2, 3, \ldots, J - 2 \end{aligned} \tag{6.103}$$

$$\begin{aligned} b_{J-1}(p + (1 - p)\sigma) = {} & (b_{J-2}p - S_{J-2})(R + 1)\sigma \\ & + b_{J-2}(1 - p)\sigma + S_{J-1}(R + 1)\sigma \\ & + (b_{J-1}p - S_{J-1})(1 - (R + 1)\sigma) + S_J \end{aligned} \tag{6.104}$$

$$b_J p = (b_{J-1}p - S_{J-1}(R + 1)\sigma + b_{J-1}(1 - p)\sigma + b_J p - S_J \tag{6.105}$$

We have not used the equation for node 1 since this will be linearly dependent on the others. Instead, we use the constraint equation

$$b_0 + \sum_{i=1}^{J} \left[ b_i + \sum_{j=1}^{R} (t_{ij} + c_{ij}) \right] = N \tag{6.106}$$

Equations (6.100) to (6.106) thus give $J(2R + 1) + 1$ independent equations that can be solved to obtain an equilibrium point, as follows.

The constraint equation (6.106), plus equations (6.97), (6.101) and (6.102) can be used to rewrite equation (6.100) as

$$(N - M(1 + Rp))\sigma = S_1(1 - (R + 1)\sigma) \tag{6.107}$$

Equation (6.103) can be rearranged as

$$\begin{aligned} & (b_{i-1}(1 + Rp) - S_{i-1}(R + 1))\sigma - S_i(1 - (R + 1)\sigma) \\ & \quad = (b_i(1 + Rp) - S_i(R + 1))\sigma - S_{i+1}(1 - (R + 1)\sigma) \\ & \qquad i = 2, 3, \ldots, J - 2 \end{aligned} \tag{6.108}$$

Writing

$$f_i = (b_{i-1}(1 + Rp) - S_{i-1}(R + 1))\sigma - S_i(1 - (R + 1)\sigma) \qquad i = 2, 3, \ldots, J - 2 \tag{6.109}$$

then equation (6.108) becomes

$$f_i = f_{i+1} \qquad i = 2, 3, \ldots, J - 2 \tag{6.110}$$

Equation (6.104) can be rearranged as

$$\begin{aligned} & (b_{J-2}(1 + Rp) - S_{J-2}(R + 1))\sigma - S_{J-1}(1 - (R + 1)\sigma) \\ & \quad = (b_{J-1}(1 + Rp) - S_{J-1}(R + 1))\sigma - S_J \end{aligned} \tag{6.111}$$

Defining

$$f_J = (b_{J-1}(1 + Rp) - S_{J-1}(R + 1))\sigma - S_J \tag{6.112}$$

then equation (6.111) becomes

$$f_{J-1} = f_J \tag{6.113}$$

Equation (6.105) can be rearranged as

$$(b_{J-1}(1 + Rp) - S_{J-1}(R + 1))\sigma - S_J = 0 \tag{6.114}$$

Clearly, we have

$$f_J = 0 \tag{6.115}$$

Then from equations (6.110), (6.113) and (6.115)

$$f_i = 0 \qquad i = 2, 3, \ldots, J \tag{6.116}$$

Substitution of equation (6.98) in (6.116) gives

$$b_i = b_{i-1} \frac{[(1 + Rp) - p(1 - p)^{M-1}(R + 1)]\sigma}{p(1 - p)^{M-1}(1 - (R + 1)\sigma)} \tag{6.117}$$

Writing

$$A = \frac{[(1 + Rp) - p(1 - p)^{M-1}(R + 1)]\sigma}{p(1 - p)^{M-1}(1 - (R + 1)\sigma)} \tag{6.118}$$

then

$$b_i = b_{i-1} A \tag{6.119}$$

Using this recursively

$$b_i = b_1 A^{i-1} \qquad i = 2, 3, \ldots, J - 1 \tag{6.120}$$

Equation (6.98) in (6.115) in a similar way gives

$$b_J = b_1 A^{J-1}(1 - (R + 1)\sigma) \tag{6.121}$$

Using equations (6.120) and (6.121) in equation (6.97) we get

$$b_1 = \frac{M(1 - A)}{1 - (R + 1)\sigma A^{J-1} - (1 - (R + 1)\sigma)A^J} \tag{6.122}$$

Finally, using this and (6.98) in (6.107)

$$(N - M(1 + Rp))\sigma = \frac{M(1 - A)p(1 - p)^{M-1}(1 - (R + 1)\sigma)}{1 - (R + 1)\sigma A^{J-1} - (1 - (R + 1)\sigma)A^J} \tag{6.123}$$

Clearly, this can be solved as a fixed point equation for $M$, to obtain an equilibrium value $M_e$. Mean throughput is then approximated using equation (6.99) as

$$E[S(x)] \simeq S = M_e p(1-p)^{M_e - 1} \tag{6.124}$$

Once this has been done, the mean packet delay can be found using Little's result by first calculating the mean number of packets stored in the system. Call this $\bar{n}$. Then

$$\bar{n} = \sum_{i=1}^{J} i\left[b_i + \sum_{j=1}^{R}(t_{ij} + c_{ij})\right] + b_0\sigma \tag{6.125}$$

Using equations (6.101), (6.102) and (6.106) then

$$\bar{n} = \sum_{i=1}^{J} i[b_i(1 + Rp)] + \left[N - \sum_{i=1}^{J}(b_i(1 + Rp))\right]\sigma \tag{6.126}$$

This can be evaluated using equations (6.120), (6.121), (6.122) and (6.118).

Having found $\bar{n}$, then the mean packet delay, $W$, is given by Little's result as

$$W = \frac{\bar{n}}{S} \tag{6.127}$$

When a system has buffered users, another important performance measure is the probability of buffer overflow, $B$. This is defined as previously to be the probability that a newly generated packet finds a full buffer. With the system in equilibrium, this is the probability that the buffer of a user is full. This can be approximated by

$$B = \left(b_J + \sum_{i=1}^{R}(t_{Ji} + c_{Ji})\right)\frac{1}{N} \tag{6.128}$$

Using equations (6.101), (6.102), (6.121) and (6.122) we obtain

$$B = \frac{M(1-A)A^{J-1}(1-(R+1)\sigma)(1+Rp)}{N(1-(R+1)\sigma A^{J-1} - (1-(R+1)\sigma)A^J)} \tag{6.129}$$

Substitution of the equilibrium value of $M$ in the above will thus enable $B$ to be evaluated.

Typical performance results for this model are given in Figs. 6.17 and 6.18, which show respectively the mean throughput and mean packet delay plotted against $p$ and compared with simulations for buffer sizes of $J = 3$ and $J = 10$. The other parameters are $N = 100$, $\sigma = 0.0042$ and $R = 12$. The results are seen to match well when the protocol is stable. This is a trend that was observed in the previous models, and is a well known feature of EPA. A special case of the model presented here has been analysed by Tasaka [TASA 86], where the channel has zero delay. The results presented here match those of Tasaka provided we set $R = 0$ in the relevant equations.

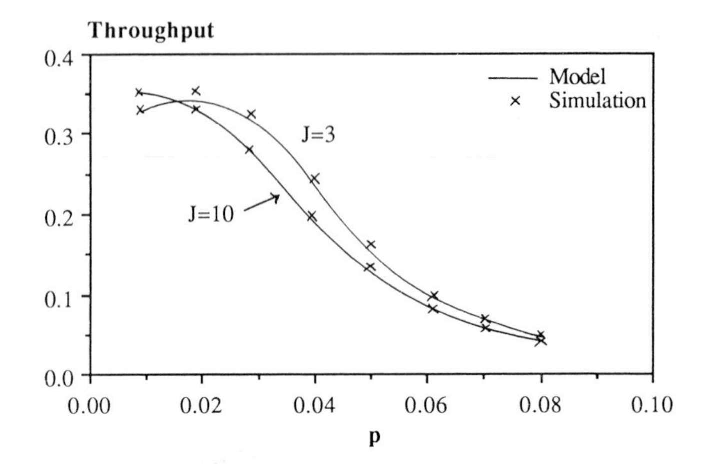


*Figure 6.17 Throughput for the buffered slotted Aloha model*

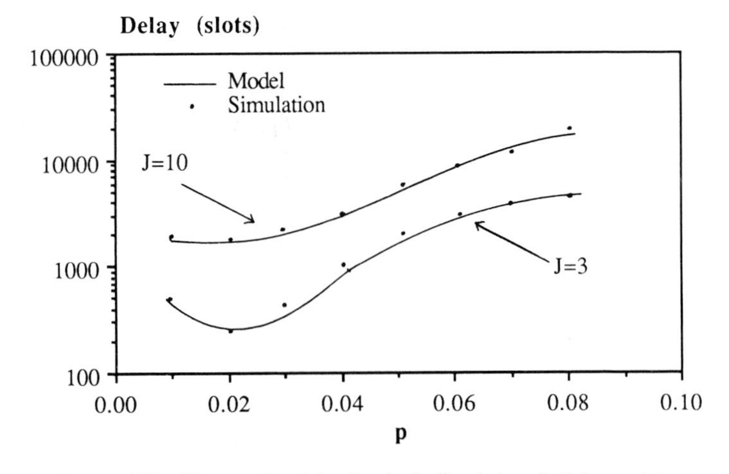


*Figure 6.18 Mean packet delay for the buffered slotted Aloha model*

## 6.5 EXERCISES

6.1 Extend the TDMA model of Section 6.1 for the case where each user has a buffer that can hold $J$ packets. Single packet messages should be assumed ($\gamma = 1$). Solve this model, and hence find expressions for the mean throughput and the mean packet delay.

6.2 Apply the EEPA method (Section 5.5.2) to evaluate the performance of the slotted Aloha model of Section 6.2.1 when each user belongs to a different customer class. This can be done in a parallel way to the REPA method in Section 6.2.2.

6.3 Investigate the possibility of applying the stochastic transformation in Section 6.2.3 to a system that has three parallel inputs to a channel rather than two; for example, slotted Aloha with buffered users and a buffer size of two (packets).

*6.4 Extend the CDMA model in Section 6.3 to the case of buffered users, single packet messages, and delayed first transmission. This should be done along the lines of the model in Section 6.4.

6.5 Modify the buffered slotted Aloha model of Section 6.4 to allow for statistically different users. This should be done using EEPA (Section 5.5.2).

# 7. LOCAL AREA NETWORKS

*Chapter Objectives: To demonstrate how various features of protocols, such as backoff strategies, polling methods and timing constraints, can be incorporated into performance models, using local networks as examples. The modelling objectives are outlined at the start of each section.*

The concept of having a number of independent users share a common transmission medium is a convenient one in computer communications. As we have seen, one example of this is a satellite network where $N$ users share a common link provided by a geostationary satellite. Such a system is characterised by the very long propagation delay on the channel, which is of the order of 0.27 seconds and thus considerably longer than the average transmission time of a packet. The ratio of channel propagation delay to mean packet transmission time is usually given the symbol $a$; thus

$$\begin{aligned} a &= \frac{\text{Channel propagation delay}}{\text{Mean packet transmission time}} \\ &= \frac{d/V}{L/C} \\ &= \frac{dC}{LV} \end{aligned} \tag{7.1}$$

where

$L$ = mean packet length (bits)
$C$ = channel bit rate (bits/s)
$d$ = length of transmission medium (m)
$V$ = signal propagation velocity (m/s)

Thus for a satellite network with a bit rate of 50 kbits/s, a mean packet length 1125 bits, and a channel delay of 0.27 seconds, we have $a = 12$. This is, of course, the channel propagation delay in units of mean packet transmission times.

If we go to the other extreme from a satellite network and consider a network within a local area, such as a university campus, a building, or room, the users can be connected by a single length of cable strung between them, usually in the form of a bus or a ring. Such networks are called *local area networks*.

For coaxial cable and other similar transmission media, $V$ is approximately two thirds the speed of light, or $2 \times 10^8$ m/s. Thus, with a transmission medium of length 1 km, packets of length 1000 bits, and a bit rate of 10 Mbits/s, we have $a = 0.05$. This is a typical value for a local area network, and implies a channel delay that is now twenty times smaller than the packet transmission time. It is this feature of a very small channel propagation delay that characterises a local area network, and can be used to advantage in such things as carrier sensing and detecting packet collisions [TANE 81]. From our point of view, when building a model of a local area network, this must be done in such a way to explicitly reflect the channel delay and so incorporate its effect on performance.

The following common assumptions will be made in the models that follow:

A7.1 The time axis is slotted with slot size $\tau$ seconds.
A7.2 The channel is noise free.
A7.3 A fixed population of $N$ users share the channel.
A7.4 There is a single user class.
A7.5 Packets are of fixed length $H$ slots ($H\tau$ seconds).

We shall broadly categorise local area networks as either *carrier sensing networks* or *polling networks*, depending on the form of the channel access mechanism that they employ. Carrier sensing networks usually have a bus architecture, whereas polling networks can have either a bus or a ring architecture, with the former arranged as a logical (rather than physical) ring.

Four types of local area networks will be considered. These are chosen to typify the techniques that can be used to model the various types of local area networks, including those not considered here. Section 7.1 deals with the modelling of *carrier sensing networks*. An important feature of such networks is the *backoff algorithm* that is used to resolve packet collisions. We thus consider how to model *carrier sense multiple access with collision detection* (CSMA/CD) networks that use *dynamic backoff* strategies. A particular example of this type of network is the IEEE 802.3 standard [IEEE 85A], commonly known as the *Ethernet* [METC 76], which uses a *binary exponential backoff algorithm*. A radically different bus accessing scheme based on a distributed form of polling is the *token bus*. An example of this type of network is the IEEE standard 802.4 [IEEE 85B], and is considered in Section 7.2.1. Also considered in the same section is another IEEE standard, the 802.5, [IEEE 85C], which is the *token ring*. As its name implies, this is a ring version of the token bus, and both fall into the category of polling networks. Extensions of the token ring in the form of the *fibre distributed data interface* (FDDI) [ROSS 89] are used in high speed, fibre optic based ring networks for transmitting integrated services. Such systems used *timed token protocols* for meeting timing deadlines on real-time services. A model for a network with a timed token protocol is considered in Section 7.2.2. The final model of a local area network is the *slotted ring*, given in Section 7.3. This might be thought of as a multiple server polling network, and although it does not form part of any international standard at the present time, it is a network that is known to perform well in the type of high speed, short packet length environments that are becoming common with the increasing use of integrated services [BHUY 89].

As in Chapter 6, the numerical results presented in this chapter are representative samples only, and are intended to give an assessment of the accuracy of the modelling methods when compared with simulations.

## 7.1 CARRIER SENSING NETWORKS

*Modelling objectives: To demonstrate how dynamic backoff schemes can be modelled.*

The modelling strategy used here typifies how the performance of dynamic backoff algorithms can be compared and assessed. A discrete-time queueing network model for a CSMA/CD network that uses dynamic backoff is shown in Fig. 7.1. In addition to the general assumptions previously given, the assumptions made in compiling this model are as follows:

(1) The channel delay is identical for all users and is $\tau = d/V$ seconds.
(2) An idle user generates a new message with probability $\sigma$ per slot.
(3) Message lengths are geometrically distributed with mean $1/\gamma$ packets.
(4) A user can have, at most, one message waiting for transmission.
(5) A user with a packet waiting for transmission senses the channel with probability $r_i$ per slot, where $i$ is the number of collisions in which the packet has been involved.
(6) If a collision is detected, the packet transmission will be terminated, and a jamming signal of length $K$ slots is transmitted to indicate to other users that a collision is taking place.

Now define the following network parameters:

$b$ = conditional probability that a user finds that the channel is busy in a particular slot, given that the channel is sensed in that slot.
$f$ = probability that a user's packet collides on a transmission attempt.

The model in Fig. 7.1 can now be interpreted as follows.

Users at node I are idle and generate a message with probability $\sigma$ per slot. Users that generate a message jump to node $B_0$ just before the end of the slot. Users at the nodes labelled $B_i$, $i = 0, 1, \ldots, m$, sense the channel with probability $r_i$ in each slot. Users that sense the channel jump from the node just before end of the slot; that is, $r_i$ is the service rate of node $B_i$. A user that senses the channel at node $B_i$ will find the channel busy with probability $b$, and so will jump back to node $B_i$. A user that senses the channel at node $B_i$ will find the channel free with probability $1 - b$. In this case one of two things can happen; the user can either be involved

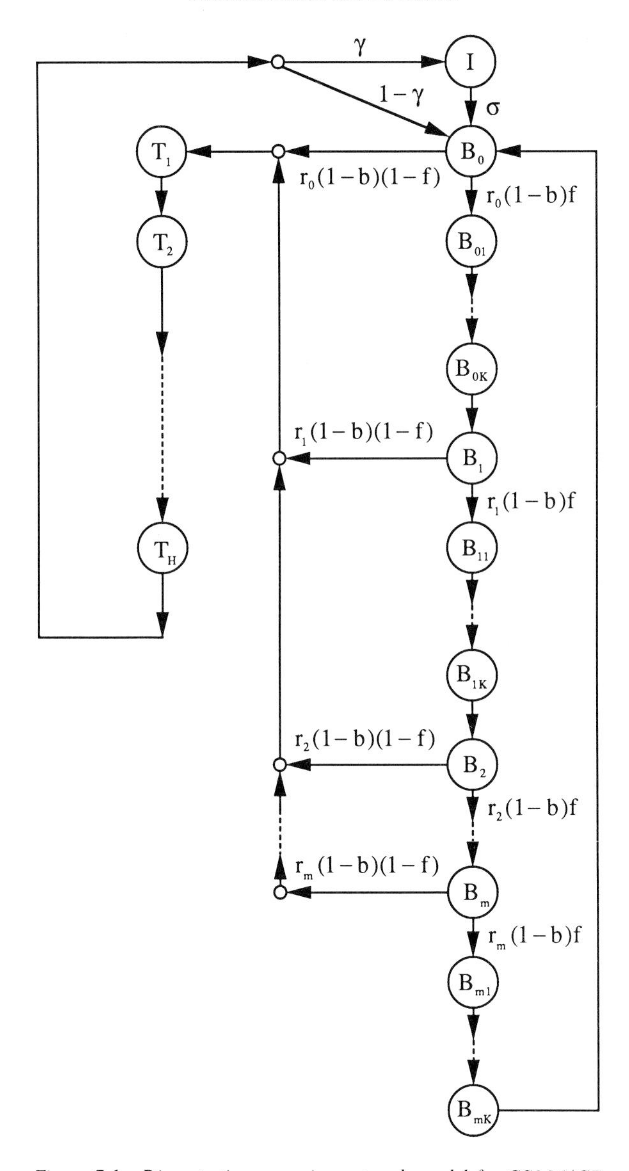


*Figure 7.1 Discrete-time queueing network model for CSMA/CD*

in a collision with probability $f$, or successfully acquire the channel with probability $1 - f$. Users that are involved in a collision jump to node $B_{i1}$ and are subject to a $K$ slot delay represented by the nodes $B_{i1}, B_{i2}, \ldots, B_{iK}$. At each of these $K$ nodes users are transmitting a jamming signal to indicate to other users that a collision is taking place. After jumping from a node $B_{iK}$ a user goes to node $B_{i+1}$, $0 \leq i \leq m-1$, or node $B_0$, $i = m$, where channel sensing again commences. A user that successfully acquires the channel at node $B_i$, $i = 0, 1, \ldots, m$, jumps to node $T_1$. In this case nodes $B_i, T_1, T_2, \ldots, T_{H-1}$ represent the $H$ slot transmission period of the successful packet.

Node $T_H$ represents the effects of the channel propagation delay. After jumping from node $T_H$ a user goes either to node I with probability $\gamma$ or node $B_0$ with probability $1-\gamma$. The former implies there are no further packets in the message and so the user again becomes idle and free to generate a new message. The latter implies there are more packets for transmission in the geometrically distributed message.

For convenience in Fig. 7.1 the arcs have been labelled directly with the transition probabilities $p_{ij}$ rather than distinguish between the service rates $\mu_i$ and the routing probabilities $r_{ij}$; that is $p_{ij} = \mu_i r_{ij}$.

The equilibrium point equations for the model of Fig. 7.1 are as follows.

$$x^I \sigma = x^{T_H} \gamma \tag{7.2}$$

$$x^{B_{i1}} = x^{B_{i2}} = \cdots = x^{B_{iK}} = x^{B_i} r_i (1-b) f; \qquad 0 \leq i \leq m \tag{7.3}$$

$$x^{T_1} = x^{T_2} = \cdots = x^{T_H} = (1-b)(1-f) \sum_{i=0}^{m} x^{B_i} r_i \tag{7.4}$$

$$x^{B_{iK}} = x^{B_{i+1}} r_{i+1} (1-b); \qquad 0 \leq i \leq m-1 \tag{7.5}$$

The normalising equation is

$$x^I + \sum_{i=1}^{H} x^{T_i} + \sum_{i=0}^{m} \left( x^{B_i} + \sum_{j=1}^{K} x^{B_{ij}} \right) = N \tag{7.6}$$

Under the assumptions made, the equation for conditional throughput at state $x$ is clearly

$$S(x) = (1-b)(1-f) \sum_{i=0}^{m} x^{B_i} r_i + \sum_{i=1}^{H-1} x^{T_i} \tag{7.7}$$

Using equation (7.4), the above can be expressed more conveniently as

$$S(x) = \sum_{i=1}^{H} x^{T_i} = H x^{T_1} \tag{7.8}$$

The idea will be to express equation (7.6) in terms of $x^{T_1}$, and then solve this to find the equilibrium value $x_e^{T_1}$. Mean throughput can then be approximated by

$$E[S(x)] \simeq S(x_e) = S = H x_e^{T_1} \tag{7.9}$$

Considering in turn each of the terms on the left hand side of equation

(7.6), using equations (7.2) and (7.4)

$$x^I = x^{T_1}\frac{\gamma}{\sigma} \tag{7.10}$$

Using equation (7.4)

$$\sum_{i=1}^{H} x^{T_i} = Hx^{T_1} \tag{7.11}$$

Note that this term is also the conditional throughput at state $x$ (equation 7.8).

We shall now make the assumption of *exponential backoff*. Other forms of backoff (e.g. linear) can be modelled by making an appropriate assumption at this point. For exponential backoff then

$$r_i = R^i \tag{7.12}$$

where $1/R$ is the *radix* of the backoff scheme. For example, for binary exponential backoff $1/R = 2$. Then from equations (7.3), (7.5) and (7.12)

$$x^{B_i} = x^{B_0}\left(\frac{f}{R}\right)^i \tag{7.13}$$

Now using equation (7.13) in (7.4)

$$x^{T_1} = x^{B_0}(1-b)(1-f)\sum_{i=0}^{m} f^i$$

$$= x^{B_0}(1-b)(1-f^{m+1})$$

Then we have

$$x^{B_0} = \frac{x^{T_1}}{(1-b)(1-f^{m+1})} \tag{7.14}$$

Using this in equation (7.13) and summing from $i = 0$ to $m$ we have

$$\sum_{i=0}^{m} x^{B_i} = \frac{x^{T_1}}{(1-b)(1-f^{m+1})}\sum_{i=0}^{m}\left(\frac{f}{R}\right)^i$$

$$= \frac{x^{T_1}}{(1-b)(1-f^{m+1})}\left(\frac{1-\left(\dfrac{f}{R}\right)^{m+1}}{1-\dfrac{f}{R}}\right) \tag{7.15}$$

To express the final term of the left hand side of equation (7.6) in terms of $x^{T_1}$ we first use equation (7.3) to write

$$\sum_{i=0}^{m}\sum_{j=1}^{K} x^{B_{ij}} = K(1-b)f\sum_{i=0}^{m} x^{B_i}R^i \tag{7.16}$$

Using equations (7.4) and (7.12), then equation (7.16) becomes

$$\sum_{i=0}^{m} \sum_{j=1}^{K} x^{B_{ij}} = x^{T_1} K \frac{f}{1-f} \tag{7.17}$$

Finally, substituting (7.10), (7.11), (7.15) and (7.17) into (7.6), the latter becomes

$$x^{T_1}\left(\frac{\gamma}{\sigma} + H + \frac{1}{(1-b)(1-f^{m+1})}\left(\frac{1-\left(\frac{f}{R}\right)^{m+1}}{1-\frac{f}{R}}\right) + \frac{Kf}{1-f}\right) = N \tag{7.18}$$

We now require to solve for the parameters $b$ and $f$. To do this, first assume that the load offered to the channel by users that are not idle, transmitting or undergoing a collision (i.e. users at nodes $B_i$, $0 \leq i \leq m$), is Poisson distributed with rate $\lambda$ packets/user/slot. Then we have

$$\lambda = \frac{\sum_{i=0}^{m} R^i x^{B_i}}{\sum_{i=0}^{m} x^{B_i}} \tag{7.19}$$

Using equations (7.15), (7.16) and (7.17), then (7.19) becomes

$$\lambda = \frac{1-f^{m+1}}{1-f} \Bigg/ \frac{1-\left(\frac{f}{R}\right)^{m+1}}{1-\frac{f}{R}} \tag{7.20}$$

Since a user that attempts a transmission will be involved in a collision if at least one of the other $N-1$ users also attempts a transmission in the same slot, then

$$f = f(\lambda) = 1 - e^{-\lambda(N-1)} \tag{7.21}$$

Equations (7.20) and (7.21) give a fixed point equation of the form

$$\lambda = \phi(\lambda) \tag{7.22}$$

Let a solution to this be $\lambda_e$, giving $f_e = f(\lambda_e)$. It now remains to find the corresponding value of $1-b$, the probability that a user (call this user the *observer*) finds the channel free upon sensing the carrier. One method of doing this is to construct a Markov chain representing the state of the channel as seen by the observer. The channel must alternate between being in an *idle state*, and in either a *transmission state* or a *collision state*, these latter, of course, being mutually exclusive. Denoting the idle state by $i$,

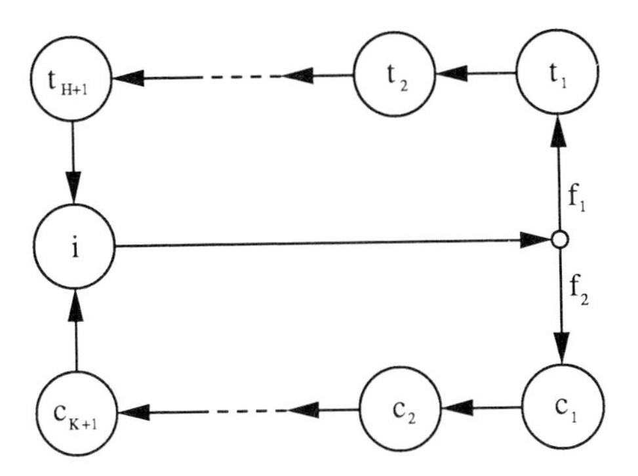


*Figure 7.2 Channel state as a Markov chain*

the transmitting states by $t_1, t_2, \ldots, t_{H+1}$ (the $H$ slot transmission period, plus one slot propagation delay before the observer detects the channel is idle again), and the collision states by $c_1, c_2, \ldots, c_{K+1}$ (a $K$ slot jamming period, followed again by one slot before an idle channel is detected), the state transition diagram of this Markov chain is shown in Fig. 7.2. The transition probabilities $f_1$ and $f_2$ denote the probability of exactly one user attempting a transmission, and two or more users attempting transmissions, respectively. Then we have

$$f_1 = e^{-\lambda(N-1)}\lambda(N-1) \tag{7.23}$$

and

$$f_2 = 1 - e^{-\lambda(N-1)} - e^{-\lambda(N-1)}\lambda(N-1) \tag{7.24}$$

Denoting the equilibrium probability that the Markov chain is in a given state $j$ as $\Pi(j)$, then the balance equations are

$$\Pi(c_1) = \Pi(c_2) = \cdots = \Pi(c_{K+1}) = f_2\Pi(i) \tag{7.25}$$

$$\Pi(t_1) = \Pi(t_2) = \cdots = \Pi(t_{H+1}) = f_1\Pi(i) \tag{7.26}$$

The normalising equation is

$$\Pi(i) + \sum_{j=1}^{H+1} \Pi(t_j) + \sum_{j=1}^{K+1} \Pi(c_j) = 1 \tag{7.27}$$

A solution to these equations gives

$$(1 - b) = \Pi(i) = \frac{1}{(K+1)f_1 + (H+1)f_2 + 1} \tag{7.28}$$

Using equations (7.23), (7.24) and (7.28), then $1 - b$ can be expressed as a function of $\lambda$, say $1 - b(\lambda)$, and hence the equilibrium value is obtained as $1 - b_e = 1 - b(\lambda_e)$ from the solution to equation (7.22). Substitution of

$f_e$ and $b_e$ in equation (7.18) thus enables the equilibrium value of $x^{T_1}$, namely $x_e^{T_1}$, to be evaluated. Throughput is then obtained from equation (7.9).

The mean message delay is obtained from Little's result as

$$W = \frac{N - x^I}{S} + \frac{x^I \sigma}{S}$$

$$= \frac{N}{S} - \frac{\gamma}{H\sigma} + \frac{\gamma}{H} \tag{7.29}$$

Note that this is in units of mean message transmission times ($H/\gamma$ slots). In units of slots, this delay becomes

$$W = \frac{NH}{S\gamma} - \frac{1}{\sigma} + 1 \tag{7.30}$$

Typical performance results obtained from this model with $N = 10$, $H = 25$, $K = \gamma = 1$, $R = \frac{1}{2}$ and $m = 8$, are shown evaluated against $10^5$ slot simulations in Figs. 7.3 and 7.4. These respectively show throughput and delay variation with message arrival probability, $\sigma$. It can be seen that agreement is, for the most part, good.

Other carrier sensing networks, such as the 1-persistent CSMA/CD with binary exponential backoff (Ethernet) can be modelled in an almost identical way to the model considered. An Ethernet model is outlined in the exercises.

Although there is a vast literature on CSMA/CD, most studies do not consider the dynamics of the backoff scheme but assume this is static (fixed). An exception is [GELE 82], in which a continuous-time version of an Ethernet model can be found.

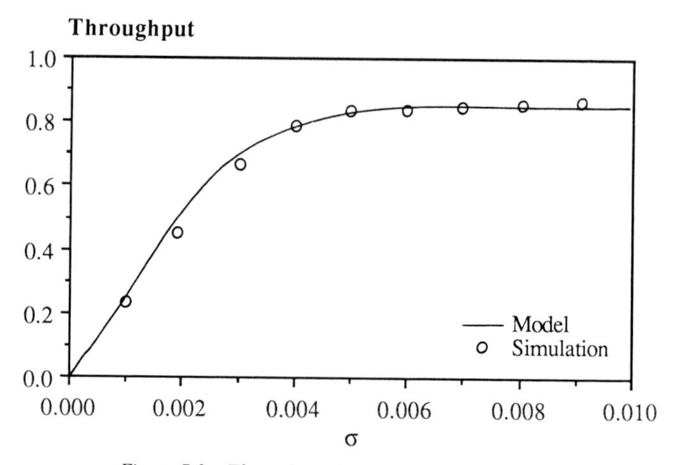


*Figure 7.3 Throughput for the CSMA/CD model*

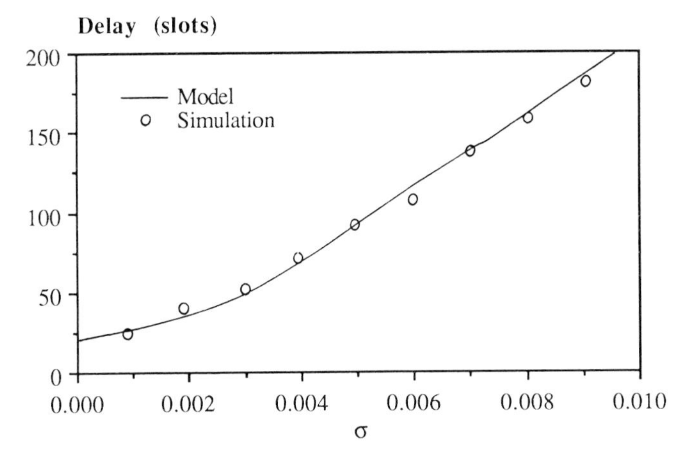


*Figure 7.4 Mean message delay for the CSMA/CD model*

## 7.2 TOKEN PASSING NETWORKS

### 7.2.1 Token bus and token ring

*Modelling objectives: To demonstrate how an accurate model for a token passing network can be developed, and show how to solve this using EPA.*

There are two major token passing local area networks, the token bus and the token ring. The *token bus* is a network that uses a very different channel access mechanism than that used in CSMA/CD, although the architectures are the same. The basic principle is that a user can only transmit on to the bus if it has permission. This permission is obtained by the user receiving a *token*, which is a special reserved bit pattern. Once a user has transmitted its data then the token is regenerated and addressed to the next user according to some pre-defined ordering. In this way permission to transmit is passed around all the active users in turn, and then the cycle is repeated.

There are a number of variants on this basic access mechanism, such as the amount of data a user can transmit upon receipt of the token. In what follows we shall use the following assumptions in deriving a discrete-time queueing network model.

(1) The channel delay is identical for all users and is $\tau = d/V$ seconds.
(2) An idle user generates a new message at the start of a slot with probability $\sigma$.
(3) Message lengths are geometrically distributed with a mean $1/\gamma$ packets.
(4) A user can have, at most, one message waiting for transmission.
(5) Upon receiving the token, a user transmits *all* packets of its message (exhaustive service).
(6) Token length is $R$ slots.

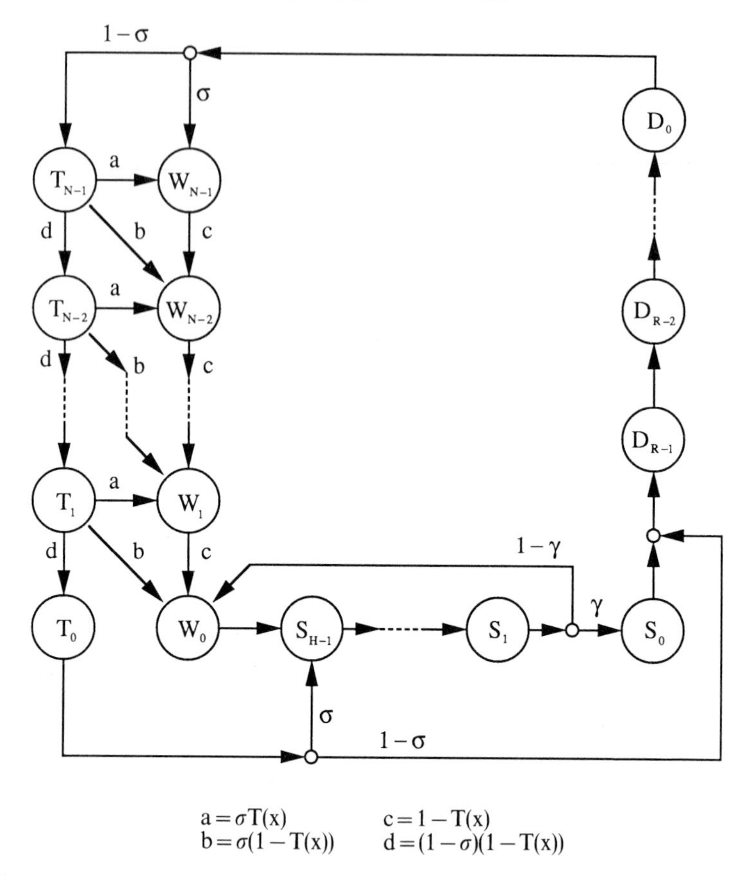


*Figure 7.5 Discrete-time queueing network model for a token bus*

In addition to the above, the global assumptions for local networks, A7.1–A7.5, apply.

Defining $T(x)$ to be the conditional probability that a user is transmitting a packet or the token, given that the system is at state $x$, then Fig. 7.5 shows a discrete-time queueing network model for a token bus under the above assumptions. In this model the service rate at all the nodes is 1, and the arcs are labelled with the routing probabilities. Note that some of these latter are functions of $T(x)$, and so are state dependent.

The model can be interpreated as follows. A user at node $T_i$, $1 \leq i \leq N - 1$, has no messages waiting, and there are $i$ users ahead of it in the queue for the token (including the user that currently has the token). This user jumps to node $W_i$ at the end of the slot if it generates a message and the user currently holding the token retains it. This event has probability $\sigma T(x)$. The user jumps to node $W_{i-1}$ at the end of the slot if it generates a message and the user currently holding the token gives it up. This event has probability $\sigma(1 - T(x))$. The user jumps to node $T_{i-1}$ if it does not generate a message and the user holding the token gives it up. This event has probability $(1 - \sigma)(1 - T(x))$. A user at node $T_0$ has just acquired the token. It can generate a message with probability $\sigma$, begin to transmit a packet in the current slot, and then jump to node $S_{H-1}$. The user does not generate a message with probability $1 - \sigma$ and jumps to node $D_{R-1}$ at the end of the slot. Users at nodes $W_i$, $1 \leq i \leq N - 1$, have a message waiting and will acquire the token after a further $i$ users have had their turn. They thus jump back to node $W_i$ or jump to node $W_{i-1}$ with probabilities $T(x)$ and $1 - T(x)$ respectively. A user at node $W_0$ has just acquired the token, and since it has a message waiting begins to transmit the first packet of this in the current slot, and then jumps to node $S_{H-1}$ at the end of the slot. Nodes $S_{H-1}, S_{H-2}, \ldots, S_1$, reflect transmission of the packet. A user at node $S_1$ either jumps to node $W_0$ at the end of the slot with probability $1 - \gamma$ or jumps to the node $S_0$ with probability $\gamma$. The former implies there are more packets in the geometrically distributed message, and the latter implies that the message transmission is complete. Node $S_0$ represents the first slot of the token transmission for users that have just completed transmitting their message (note that node $T_0$ represents the same thing for users that have no messages to transmit and do not generate a message in the current slot). Nodes $D_{R-1}, D_{R-2}, \ldots, D_1$, represent the remainder of the token transmission time, and node $D_0$ represents the effects of the channel propagation delay. Note that a user at node $D_0$ is free to generate a new message in the current slot.

The equilibrium point equations for this model are:

$$x^{T_{N-1}}(1 - (1 - \sigma)T(x)) = x^{D_0}(1 - \sigma) \tag{7.31}$$

$$x^{W_{N-1}}(1 - T(x)) = x^{T_{N-1}}\sigma T(x) + d^{D_0}\sigma \tag{7.32}$$

$$x^{T_i}(1 - (1 - \sigma)T(x)) = x^{T_{i+1}}(1 - \sigma)(1 - T(x)), \qquad 1 \leq i \leq N - 2 \tag{7.33}$$

$$x^{W_i}(1 - T(x)) = x^{W_{i+1}}(1 - T(x)) + x^{T_{i+1}}\sigma(1 - T(x)) + x^{T_i}\sigma T(x), \qquad 1 \leq i \leq N - 2 \tag{7.34}$$

$$x^{T_0} = x^{T_1}(1 - \sigma)(1 - T(x)) \tag{7.35}$$

$$x^{W_0} = x^{T_1}\sigma(1 - T(x)) + x^{W_1}(1 - T(x)) + x^{S_1}(1 - \gamma) \tag{7.36}$$

$$x^{S_{H-1}} = x^{S_{H-2}} = \cdots = x^{S_1} = x^{W_0} + x^{T_0}\sigma \tag{7.37}$$

$$x^{S_0} = x^{S_1}\gamma \tag{7.38}$$

$$x^{D_{R-1}} = x^{D_{R-2}} = \cdots = x^{D_0} = x^{S_0} + x^{T_0}(1 - \sigma) \tag{7.39}$$

In addition to the above equations, we have the following constraint on the state space:

$$\sum_{i=0}^{N-1} (x^{T_i} + x^{W_i}) + \sum_{i=0}^{H-1} x^{S_i} + \sum_{i=0}^{R-1} x^{D_i} = N \tag{7.40}$$

More specifically, this can be expressed by the following equations:

$$\sum_{i=0}^{H-1} x^{S_i} + \sum_{i=0}^{R-1} x^{D_i} + x^{T_0} + x^{W_0} = 1 \tag{7.41}$$

$$x^{T_i} + x^{W_i} = 1, \qquad 1 \leq i \leq N - 1 \tag{7.42}$$

The objective is now to solve these equations for an equilibrium point by first reducing them to a fixed point equation for one of the variables (say $x^{T_0}$). To this end we shall first find $T(x)$ as a function of $x^{T_0}$.

By definition of $T(x)$, this can be expressed as

$$T(x) = x^{T_0} + x^{W_0} + \sum_{i=0}^{H-1} x^{S_i} + \sum_{i=1}^{R-1} x^{D_i} \tag{7.43}$$

Using equation (7.35)

$$x^{T_1} = \frac{x^{T_0}}{(1 - \sigma)(1 - T(x))} \tag{7.44}$$

From equation (7.42)

$$x^{W_1} = 1 - x^{T_1} \tag{7.45}$$

Substituting (7.44) and (7.45) in (7.36) then

$$x^{W_0} = \frac{x^{T_0}}{(1 - \sigma)(1 - T(x))}\sigma(1 - T(x)) + \left[1 - \frac{x^{T_0}}{(1 - \sigma)(1 - T(x))}\right](1 - T(x)) + x^{S_1}(1 - \gamma) \tag{7.46}$$

Substituting for $x^{S_1}$ from equation (7.37), then equation (7.46) gives

$$x^{W_0} = x^{T_0}\left[\frac{\sigma(1 - \gamma) - 1}{\gamma}\right] + \left[\frac{1 - T(x)}{\gamma}\right] \tag{7.47}$$

Now from equations (7.37) and (7.38)

$$\sum_{i=0}^{H-1} x^{S_i} = (H - 1 + \gamma)(x^{W_0} + x^{T_0}\sigma) \tag{7.48}$$

From equations (7.39)

$$\sum_{i=1}^{R-1} x^{D_i} = R(x^{S_0} + x^{T_0}(1 - \sigma)) - x^{D_0} \tag{7.49}$$

Also, equations (7.37) and (7.38) give

$$x^{S_0} = \gamma(x^{W_0} + x^{T_0}\sigma) \tag{7.50}$$

Using this in equation (7.39) gives

$$x^{D_0} = \gamma(x^{W_0} + x^{T_0}\sigma) + x^{T_0}(1 - \sigma) \tag{7.51}$$

Substituting (7.50) and (7.51) into equation (7.49) then

$$\sum_{i=1}^{R-1} x^{D_i} = (R - 1)[\gamma(x^{W_0} + x^{T_0}\sigma) + x^{T_0}(1 - \sigma)] \tag{7.52}$$

We are now in a position to evaluate $T(x)$. Substituting (7.47), (7.48) and (7.52) into equation (7.43), the latter becomes

$$T(x) = \frac{\gamma R + H(1 - x^{T_0}(1 - \sigma))}{\gamma(1 + R) + H} \tag{7.53}$$

We now require an independent equation for $x^{T_0}$ as a function of $T(x)$. Using equation (7.33) recursively, then

$$x^{T_{N-1}} = x^{T_1}\left[\frac{1 - (1 - \sigma)T(x)}{(1 - \sigma)(1 - T(x))}\right]^{N-2} \tag{7.54}$$

From equation (7.35)

$$x^{T_1} = \frac{x^{T_0}}{(1 - \sigma)(1 - T(x))} \tag{7.55}$$

Using (7.55) in equation (7.54) then

$$x^{T_{N-1}} = x^{T_0}\frac{[1 - (1 - \sigma)T(x)]^{N-2}}{[(1 - \sigma)(1 - T(x))]^{N-1}} \tag{7.56}$$

Now bringing in the constraint equation (7.42)

$$x^{T_{N-1}} = 1 - x^{W_{N-1}} \tag{7.57}$$

and from equation (7.32)

$$x^{W_{N-1}} = \frac{x^{T_{N-1}}\sigma T(x) + d^{D_0}\sigma}{1 - T(x)} \tag{7.58}$$

Using (7.57) and (7.31) in equation (7.58) we get

$$x^{T_{N-1}} = \frac{(1 - \sigma)(1 - T(x))}{1 - (1 - \sigma)T(x)} \tag{7.59}$$

Equating (7.56) and (7.59) then

$$x^{T_0} = (1 - \sigma)(1 - T(x))\left[\frac{(1 - \sigma)(1 - T(x))}{1 - (1 - \sigma)T(x)}\right]^{N-1} \tag{7.60}$$

Substituting equation (7.53) into equation (7.60) turns the latter into the required fixed point equation of the form

$$x^{T_0} = f(x^{T_0})$$

This can be solved numerically to obtain $x^{T_0}$ at an equilibrium point.

To derive the performance measures, first note that from equations (7.39), (7.41) and (7.43)

$$T(x) = 1 - x^{T_0}(1 - \sigma) - x^{S_0} \tag{7.61}$$

and that at state $x$ conditional throughput is

$$S(x) = x^{T_0}\sigma + x^{W_0} + \sum_{i=1}^{H-1} x^{S_i} \tag{7.62}$$

Substituting (7.61) and (7.62) into equation (7.41) then

$$S(x) = T(x) - \sum_{i=0}^{R-1} x^{D_i} \tag{7.63}$$

From equations (7.39) and (7.61) we have

$$\sum_{i=0}^{R-1} x^{D_i} = R[1 - T(x)] \tag{7.64}$$

hence from equations (7.63) and (7.64)

$$S(x) = T(x)(1 + R) - R \tag{7.65}$$

Mean throughput is then obtained by substituting the equilibrium value, $x_e$, in the above, with $T(x_e)$ obtained from equation (7.53).

To compute mean message delay, let $\bar{n}$ be the number of users at nodes $W_i$, $i = 0, 1, \ldots, N - 1$, and $S_j$, $j = 0, 1, \ldots, H - 1$, with the network at an equilibrium point $x_e$.

Then we must have

$$S(x_e) = \left(N - \bar{n} - \sum_{i=0}^{R-1} x_e^{D_i}\right)\sigma \tag{7.66}$$

Using equations (7.51) and (7.61), (7.66) becomes

$$S(x_e) = \frac{[N - \bar{n} - R\{1 - T(x)\}]\sigma H}{\gamma} \tag{7.67}$$

From this $\bar{n}$ can be extracted as

$$\bar{n} = N - S(x_e)\frac{\gamma}{\sigma H} - R[1 - T(x_e)] \tag{7.68}$$

Using Little's result, mean message delay, $W$, is

$$W = \frac{\bar{n}}{S(x_e)} + \frac{\gamma}{H} \tag{7.69}$$

The $\gamma/H$ term on the right hand side is due to messages being generated at the start of a slot. The mean number of messages stored in the system must therefore be augmented by the mean number of users at nodes $T_i$, $i = 0, 1, \ldots, N-1$, that generate messages in a slot. This number is precisely equal to $S(x_e)\gamma/H$.

Using equations (7.68) and (7.69) we have

$$W = \frac{N}{S(x_e)} - \frac{\gamma}{\sigma H} - \frac{R[1 - T(x_e)]}{S(x_e)} + \frac{\gamma}{H} \tag{7.70}$$

This expression has units of mean message transmission times (H/$\gamma$ slots). In units of slots $W$ becomes

$$W = \frac{NH}{S(x_e)\gamma} - \frac{1}{\sigma} - \frac{HR[1 - T(x_e)]}{\sigma\gamma} + 1 \tag{7.71}$$

The performance of this model is shown assessed against $10^5$ slot simulations in Figs. 7.6 and 7.7 which show, respectively, the variation of throughput and mean message delay with message arrival probability $\sigma$ for a system with $N=10$, $H=25$ and $R=\gamma=1$. The agreement is excellent throughout.

The *token ring* can be modelled in an identical way to the above by making the assumptions that:

(1) Users are equally spaced around the ring.
(2) A slot is $1/N$ of the total time for a signal to circumnavigate the ring ($1/N$ of the ring latency)

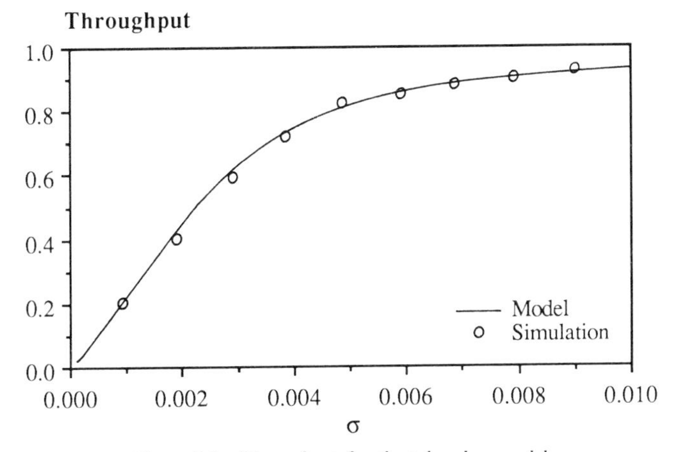


*Figure 7.6 Throughput for the token bus model*

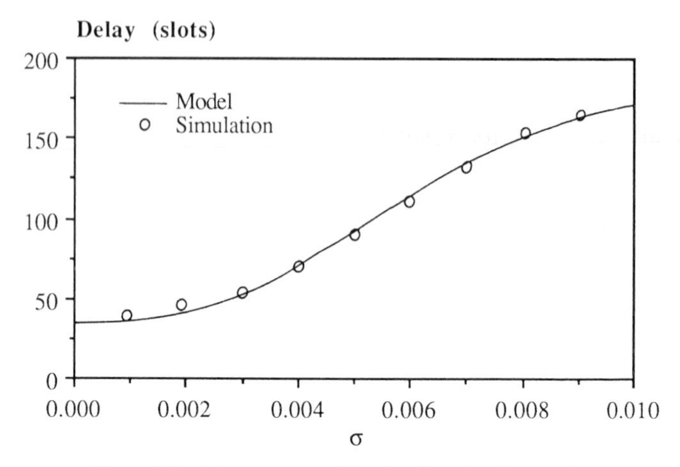


*Figure 7.7 Mean message delay for the token bus model*

A little thought will show that the token bus model carries through directly to the token ring provided the above assumptions are made. In the case of the token ring, node $D_0$ again represents the effects of the channel propagation delay in the sense that a user at this node has just released the token which is in the process of propagating to the next downstream user.

Note that the model examined represents an early token release system, such as the FDDI [ROSS 89], where the token can be considered as part of a frame header.

Alternative discrete-time models for similar systems can be found in [TAKA 86], [BOXM 88] and [SCHW 87]. Also, Tasaka has used EPA to analyse an implicit token passing scheme (fair BRAM) [TASA 86].

### 7.2.2 Timed token protocols

*Modelling objectives: To show how timing constraints (as in real-time services) can be modelled. Also, to show how a numerical solution can be effected when the EPA equations cannot be solved explicitly.*

When networks are used to provide real-time services, then the protocols are required to meet *time deadlines.* An example of this can be found in the *time-controlled token-passing protocol* of the FDDI [ROSS 86], [ROSS 89]. In such protocols, the reception of a free token by a user does not necessarily imply the transmission of a message by that user. In FDDI a *lateness counter* at each user is incremented if the token does not arrive within a *target token rotation time* (TTRT) after the token's arrival in the preceding cycle. If the token arrives within the TTRT and the lateness counter is not on zero, the counter is decremented, but messages of low priorities are not transmitted. If the lateness counter is on zero when the token arrives within the TTRT, low priority messages can be transmitted until the *token holding time* (THT) expires.

A model for this type of operation can be developed by extending the model of the token ring given in the previous section (Fig. 7.5). Before doing this however, we specify precisely the operation of the protocol to be modelled.

In addition to the assumptions A7.1–A7.5, the following will also apply:

(1) The system has a ring topology with users equally spaced around the ring.
(2) A slot is $1/N$ of the ring latency ($\tau$ seconds).
(3) The lateness in the token rotation is considered for the preceding cycle only.
(4) A user generates a message in a slot with probability $\sigma$.
(5) There are two classes of message: class 1 and class 2, class 1 having the higher priority.
(6) A generated message has probability $\sigma_1$ of belonging to class 1, and probability $\sigma_2$ of belonging to class 2, where $\sigma_1 + \sigma_2 = 1$.
(7) Messages consist of single packets of fixed length $H$ slots.
(8) Token length is $R$ slots.
(9) A buffer is emptied when a message transmission is complete.
(10) Any messages that arrive while a buffer is occupied are lost.

Define the *token rotation time* (TRT) for a user as the time between two successive instants when a free token arrives at that user. If a free token arrives at a user that has a class 1 message waiting, this is *always* transmitted. If a free token arrives at a user that has a class 2 message waiting, this is transmitted only if the preceding TRT for that user does not exceed the TTRT. The TTRT is set to a value of $N(R + 1) + MH$. The first component of this corresponds to each of the $N$ users transmitting the token to the next downstream user; that is, each user takes $R$ slots to transmit the token, plus one slot for the token to propagate to the next user. The second component of the TTRT consists of the time to transmit $M$ packets, each of length $H$ slots. $M$ is an integer chosen such that $0 \leq M \leq N$. When $M = N$ then there is no distinction between class 1 and class 2 messages; messages of either class will, in this case, always be transmitted when a free token arrives. When $M = 0$, then class 2 messages can only be transmitted if there were no transmissions at all in the preceding TRT.

Figure 7.8 shows a discrete-time queueing network model for a timed token protocol of the type described above. This is an extension of the model for the token bus given in Fig. 7.5. This latter can now be considered as a special case of Fig. 7.8 in which the class 2 message generation process has been shut down ($\sigma_2 = 0$). The main difference in the models is that

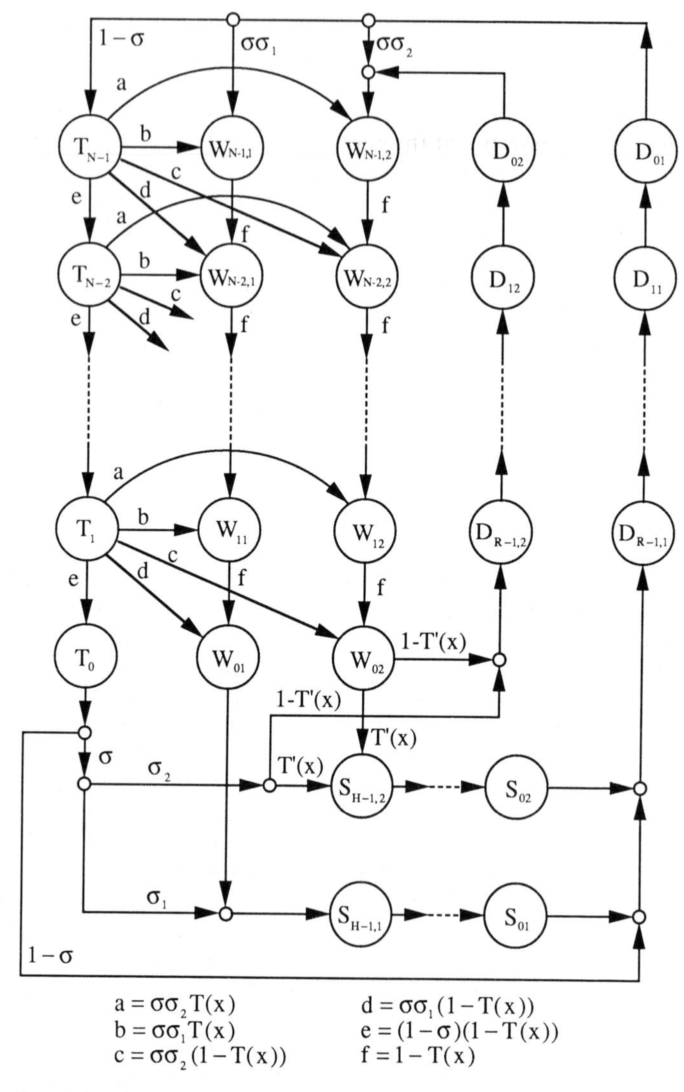


*Figure 7.8 Discrete-time queueing network model for a timed token protocol*

each of the nodes $W_i$, $i = 0, 1, \ldots, N-1$ in Fig. 7.5 has been replaced by two nodes in Fig. 7.8 labelled $W_{i,1}$ and $W_{i,2}$, $i = 0, 1, \ldots, N-1$. Users at these nodes have class 1 or class 2 messages waiting, respectively. In a similar way each of the nodes $S_i$, $i = 0, 1, \ldots, H-1$ in Fig. 7.5 has been replaced by two nodes $S_{i,1}$ and $S_{i,2}$, $i = 0, 1, \ldots, H-1$. Users at these nodes are transmitting class 1 messages or class 2 messages, respectively.

A user at node $T_i$, $i = 0, 1, \ldots, N - 1$ has an empty buffer and can generate a class $j$ message with probability $\sigma\sigma_j$, $j = 1, 2$. If a class $j$ message, $j = 1, 2$, is generated by a user at node $T_i$, $i = 1, 2, \ldots, N - 1$, and the user currently holding the token does not give this up, then the user at node $T_i$ will jump to node $W_{i,j}$ at the end of the slot. This event has probability $\sigma\sigma_j T(x)$, where $T(x)$ has its previous meaning. If a class $j$ message is generated by a user at node $T_i$, $i = 1, 2, \ldots, N - 1$, and the user currently holding the token gives this up, the user at node $T_i$ will jump to node $W_{i-1,j}$ at the end of the slot. This event has probability $\sigma\sigma_j(1 - T(x))$. If no message is generated by a user at node $T_i$, $i = 1, 2, \ldots, N - 1$ and the user currently holding the token gives this up, then the user at node $T_i$ will jump to node $T_{i-1}$ at the end of the slot. This event has probability $(1 - \sigma)(1 - T(x))$. A user at node $W_{i,j}$, $i = 1, 2, \ldots, N - 1$, $j = 1, 2$, will jump to node $W_{i-1,j}$ at the end of the slot if the user currently holding the token gives this up. This event has probability $1 - T(x)$. A user at nodes $T_0$, $W_{01}$ or $W_{02}$ has just received the token at the start of the current slot. If a user at node $T_0$ does not generate a message then it simply transmits the token in the current slot, and jumps to node $D_{R-1,1}$ at the end of the slot. The nodes $D_{R-1,1}, D_{R-2,1}, \ldots, D_{11}$ represent the token transmission, and node $D_{01}$ the channel propagation delay as before. A user at node $W_{01}$ transmits its class 1 message in the current slot and then jumps to node $S_{H-1,1}$.

This is also the case for a user at node $T_0$ that generates a class 1 message in the current slot (probability $\sigma\sigma_1$). Nodes $S_{H-1,1}, S_{H-2,1}, \ldots, S_{11}$ represent a class 1 message transmission, with $S_{01}$ the first slot of the token transmission. A user at node $W_{02}$ has a class 2 message waiting and this can be transmitted if and only if the current TRT is not greater than the TTRT. This has probability $T'(x)$ (to be defined). If the TRT > TTRT (probability $1 - T'(x)$), then the class 2 message is not transmitted but is retained in the buffer. In this case the user jumps to node $D_{R-1,2}$, and traverses the sequence of nodes $D_{R-1,2}, D_{R-2,2}, \ldots, D_{02}$, which represent passing the token. Finally the user jumps to node $W_{N-1,2}$ and joins the end of the queue of users waiting to transmit their messages. Any user at node $T_0$ that generates a class 2 message will also traverse a similar sequence of nodes to that just described for a user at node $W_{02}$.

Clearly, the key issue here is how to calculate the state-dependent routing probability $T'(x)$ that controls transmission of the class 2 messages; that is, before we can find an equilibrium point $x_e$ we must find an equation for $T'(x_e)$. To do this first define $T'(x)$ as

$$T'(x) \equiv \begin{cases} 1 & \textit{if the } TRT \leq TTRT \textit{ at state } x \\ 0 & \textit{otherwise} \end{cases} \tag{7.72}$$

Then the expectation $E[T'(x)]$ with respect to $x$ is given by

$$E[T'(x)] = P(T'(x) = 1) \tag{7.73}$$

Since the expectation of a random variable can be approximated by its value at an equilibrium point, then,

$$E[T'(x)] \simeq T'(x_e) \tag{7.74}$$

Thus $T'(x_e)$ can be approximated by $P(T'(x) = 1)$.

To find $P(T'(x) = 1)$, let $P(x)$ be the conditional probability that when the token arrives at a user, it finds in the buffer either a class 1 message or a class 2 message *that will be transmitted*, given the state is $x$.

Noting that the constraints on the state space in this model are:

$$x^{T_i} + \sum_{j=1}^{2} x^{W_{ij}} = 1, \qquad 1 \leq i \leq N - 1 \tag{7.75}$$

and

$$x^{T_0} + \sum_{j=1}^{2} \left( x^{W_{0J}} + \sum_{i=0}^{H-1} x^{S_{ij}} + \sum_{i=0}^{R-1} x^{D_{ij}} \right) = 1 \tag{7.76}$$

then clearly in the slot that the token arrives at a user, equation (7.76) must reduce to

$$x^{T_0} + \sum_{j=1}^{2} x^{W_{0j}} = 1 \tag{7.77}$$

Conditioning on this slot then

$$P(x) = \frac{x^{T_0}\sigma\sigma_1 + x^{W_{01}} + (x^{T_0}\sigma\sigma_2 + x^{W_{02}})T'(x)}{x^{T_0} + x^{W_{01}} + x^{W_{02}}} \tag{7.78}$$

Then with the system in equilibrium and assuming users are independent, the probability that the token encounters $M$ or fewer messages that will be transmitted during a complete cycle of all $N$ users gives

$$T'(x_e) = \sum_{k=0}^{M} \binom{N}{k} P(x_e)^k (1 - P(x_e))^{N-k} \qquad 0 \leq M \leq N \tag{7.79}$$

The state dependent routing probability $T(x)$ can be obtained from a direct extension of equation (7.43) as

$$T(x) = x^{T_0} + \sum_{j=1}^{2} \left( x^{W_{0j}} + \sum_{i=0}^{H-1} x^{S_{ij}} + \sum_{i=1}^{R-1} x^{D_{ij}} \right) \tag{7.80}$$

At this point one could set up the EPA equations and attempt to solve these in the usual way. In this particular case, this would prove an extremely tedious and laborious task, and it is easier to proceed numerically. This alternative approach is considered here.

Once all the transition probabilities in the model are known, then the nodes can be suitably labelled in one dimension as $i = 1, 2, \ldots, G$, where $G$ is the total number of nodes. A transition probability matrix $P$ can then be defined, where $P = [p_{ij}, i, j = 1, 2, \ldots, G]$ and the $p_{ij}$ are the node transition probabilities. If the state (vector) of the system is $x$, then the idea is to solve the set of nonlinear equations

$$x = xP \tag{7.81}$$

where

$$x = (x^{T_i}, x^{W_{ij}}, x^{S_{kj}}, x^{D_{mj}}, 0 \leq i \leq N-1, 1 \leq j \leq 2, 0 \leq k \leq H-1, \; 0 \leq m \leq R-1) \tag{7.82}$$

A number of methods can be used for this, but due to the sparsity of the transition matrix it is usually better to use an iterative method that preserves this sparsity pattern, such as the power method [KING 90]. This consists of starting with some initial estimate for $x$, say $x(0)$, that satisfies the state space constraints, and then repeatedly post multiplying the current estimate of the state vector $x(i)$ by the current estimate of the transition matrix $P(i)$, until the sequence of vectors

$$x(i+1) = x(i)P(i) \qquad i = 0, 1, 2, \ldots \tag{7.83}$$

converges to some suitable accuracy.

Note that the transition matrix itself must be updated during this process, since it contains state-dependent probabilities. Also, it is interesting to note that in the context of the present example, equations (7.78) and (7.79) show that equation (7.79) is a fixed point equation of the form

$$T'(x) = f(T'(x)) \tag{7.84}$$

Then at each iteration step of the power method the above fixed point equation must be solved to find the new value of $T'(x)$. Also, at each iteration step the new state vector must be normalised to satisfy the constraints on the state space (equations 7.75 and 7.76).

Note also that equation (7.81) is itself, in effect, a fixed point equation of the form $x = \phi(x)$, with $x$ now a vector. We thus have one fixed point equation imbedded in another.

This numerical form of EPA requires, in total, the iterative solution of $3N + 2(R + H)$ non-linear equations. This is a considerable improvement on other similar existing models for timed token protocols, such as that by Takagi [TAKA 90] for example. This latter requires the solution of $4^N$ linear equations, which makes the model impossible to use for all but small values of $N$.

The disadvantage with the use of multivariate iterative methods is that, unlike methods such as bisection that can be used with single variable problems, convergence is not guaranteed. For example, if the general form of the non-linear equations (7.81) are expanded we have a set of $G$ non-linear equations of the form

$$x^1 = f_1(x^1, x^2, \ldots, x^G)$$

$$x^2 = f_2(x^1, x^2, \ldots, x^G)$$
$$\vdots \qquad \vdots$$
$$x^G = f_G(x^1, x^2, \ldots, x^G)$$

Then it can be shown that an iterative solution can be obtained for these provided that

$$\sum_{i=1}^{G} \left| \frac{\partial f_i}{\partial x^i} \right| < 1$$

for all $j = 1, 2, \ldots, G$. (See, for example, [ORTE 70] and [BAKE 81].) Clearly, this will restrict the usefulness of the method, and some alternative solution technique may have to be sought if the method does not converge.

Once an equilibrium point $x_e$ has been found, the performance measures can easily be obtained. Throughput is given by

$$S(x_e) = \sum_{i=0}^{H-1} \sum_{j=1}^{2} x_e^{S_{ij}} \tag{7.85}$$

with class $j$ throughput given as

$$S_j = S_j(x_e) = \sum_{i=0}^{H-1} x_e^{S_{ij}}, \qquad j = 1, 2 \tag{7.86}$$

The mean message delay for class 1 and class 2 can be obtained in the usual way from Little's result by first calculating the mean number of class 1 and class 2 messages in the system. Let these be given by $\bar{n}_1$ and $\bar{n}_2$, respectively. Then we have

$$\bar{n}_1 = \sum_{i=0}^{N-1} (x^{T_i}\sigma\sigma_1 + x^{W_{i1}}) + \sum_{i=1}^{H-1} x^{S_{i1}} \tag{7.87}$$

$$\bar{n}_2 = \sum_{i=0}^{N-1} (x^{T_i}\sigma\sigma_2 + x^{W_{i2}}) + \sum_{i=1}^{H-1} x^{S_{i2}} + \sum_{i=0}^{R-1} x^{D_{i2}} \tag{7.88}$$

Then mean message delay for class $j$, $W_j$, is given by

$$W_j = \frac{\bar{n}_j}{S_j}, \qquad j = 1, 2 \tag{7.89}$$

The mean TRT can be obtained from Little's result by first noting that the rate of flow, $\lambda$, of users into nodes $T_0$, $W_{01}$ and $W_{02}$ (where the token arrives) from nodes $T_1$, $W_{11}$ and $W_{12}$ is simply

$$\lambda = x^{T_0} + x^{W_{01}} + x^{W_{02}} \tag{7.90}$$

**Table 7.1** THROUGHPUT FOR A TIMED TOKEN PROTOCOL

| | *Class 1 Throughput* | | | *Class 2 Throughput* | | |
|---|---|---|---|---|---|---|
| $\sigma$ | *Simulation* | *Modelling* | *% error* | *Simulation* | *Modelling* | *% error* |
| 0.0010 | 0.0080 | 0.0079 | −1.2500 | 0.024 | 0.0317 | 29.9180 |
| 0.0020 | 0.0136 | 0.0157 | 15.4412 | 0.0592 | 0.0629 | 6.2500 |
| 0.0030 | 0.0248 | 0.0234 | −5.6452 | 0.0928 | 0.0934 | 0.6466 |
| 0.0040 | 0.0344 | 0.0308 | −10.4651 | 0.1312 | 0.1234 | −5.9451 |
| 0.0050 | 0.0448 | 0.0382 | −14.7321 | 0.1508 | 0.1526 | 1.1936 |
| 0.0060 | 0.0404 | 0.0453 | 12.1287 | 0.1772 | 0.1812 | 2.2573 |
| 0.0070 | 0.0528 | 0.0523 | −0.9470 | 0.2180 | 0.2090 | −4.1284 |
| 0.0080 | 0.0604 | 0.0590 | −2.3179 | 0.2360 | 0.2360 | −0.0000 |
| 0.0090 | 0.0684 | 0.0656 | −4.0936 | 0.2756 | 0.2621 | −4.8984 |
| 0.0100 | 0.0696 | 0.0719 | 3.3046 | 0.2796 | 0.2873 | 2.7539 |
| | | Mean % error | −7.0325 | | Mean % error | −5.7991 |

**Table 7.2** MEAN MESSAGE DELAY FOR A TIMED TOKEN PROTOCOL

| | *Class 1 Delay* | | | *Class 2 Delay* | | |
|---|---|---|---|---|---|---|
| $\sigma$ | *Simulation* | *Modelling* | *% error* | *Simulation* | *Modelling* | *% error* |
| 0.0010 | 9.9000 | 10.4107 | 5.1586 | 9.8525 | 10.4108 | 5.6666 |
| 0.0020 | 9.5588 | 10.8528 | 13.5373 | 10.8528 | 10.8528 | 3.6938 |
| 0.0030 | 10.4839 | 11.3223 | 7.9970 | 10.3621 | 11.3223 | 9.2665 |
| 0.0040 | 11.0814 | 11.8229 | 6.6914 | 11.4055 | 11.8230 | 3.6605 |
| 0.0050 | 10.8036 | 12.3570 | 14.3785 | 11.1194 | 12.3571 | 11.1310 |
| 0.0060 | 11.1584 | 12.9278 | 15.8571 | 11.0181 | 12.9281 | 17.3351 |
| 0.0070 | 11.5985 | 13.5360 | 16.7047 | 11.8055 | 13.5370 | 14.7779 |
| 0.0080 | 12.4967 | 14.1482 | 13.5036 | 12.5068 | 14.1869 | 13.4335 |
| 0.0090 | 13.1404 | 14.8751 | 13.2013 | 13.3483 | 14.8817 | 11.4876 |
| 0.0100 | 12.4425 | 15.6074 | 25.4362 | 13.0014 | 15.6224 | 20.1594 |
| | | Mean % error | 13.2466 | | Mean % error | 11.0501 |

Then the mean $TRT$, $\overline{TRT}$, is given by

$$\overline{TRT} = \frac{N}{\lambda_e} \tag{7.91}$$

where $\lambda_e$ is the equilibrium value of $\lambda$.

Typical performance results obtained from this model are given in tabular form in Tables 7.1 and 7.2, which show throughput and mean message delay compared with results obtained from a $10^5$ slot simulation. The parameters used are $N = 10$, $\sigma_1 = 0.2$, $H = 4$, $R = 0$, and $M = 5$. Mean errors are around 7% and 6% for class 1 and class 2 throughput respectively. Mean values of class 1 and class 2 mean message delay errors are somewhat larger at around 13% and 11% respectively.

## 7.3 SLOTTED RINGS

*Modelling objectives: To show how multiple server polling systems having buffered users can be modelled.*

The token bus and the token ring are examples of *single server polling systems* in that the token (sever) visits each of the users in turn in some predefined order on a cyclic basis. The *slotted ring* might be considered as an extension of the token ring in the sense that it can be classed as a *multiple server polling system.* In this case the servers are the slots, which again visit each user in turn on a cyclic basis. It is a system that was independently proposed by Farber and Larson [FARB 72], and Pierce [PIER 72], and has since been developed along more commercial lines as the Cambridge Ring [NEED 82], and the Orwell Ring [FALC 85].

In its basic form the slotted ring is very simple. It consists of a ring of cable, or optical fibre, along which signals propagate in one direction only. The users are connected to the ring at various intervals, the ring then acting as a transmission medium between them. At any given time there can perhaps be several hundreds of bits in flight, travelling around the ring. The delay to support these bits is provided by the ring itself, plus the user interfaces that are inserted into the ring. The total bits travelling around the ring are divided into an integer number of slots. We shall consider that each slot is sufficiently large to transport a single, fixed length packet.

The medium access control protocol for a slotted ring that uses source release of slots functions as follows. Slots circulating around the ring are designated as either full or empty, depending on whether or not they contain a packet, respectively, with the first bit of the slot acting as a full/empty flag. Assuming that users are buffered, then those that have packets to transmit simply wait for the next empty slot to pass, and load this with the packet at the head of their queue. The packet will contain the address of the destination user, which will thus be recognised by the latter as it passes through its interface. Upon recognising its own address, the destination user will copy the packet into its memory. The slot, still marked as full, continues around the ring until it arrives back at the source user, which then marks the slot empty and passes it to the next downstream user. Any user that holds a slot on the ring is not allowed to transmit a second packet until the slot has been emptied. Nor is a user that has just emptied a slot allowed to reuse the same slot. These latter two rules of the protocol are to ensure that all users have a fair share of the system's transmission capacity.

In the foregoing it should be noted that we have ignored error recovery procedures, but errors are usually so infrequent, particularly when optical fibre waveguides are used as the transmission medium, that any performance overheads due to transmission errors can be ignored without measurably affecting performance results in any way.

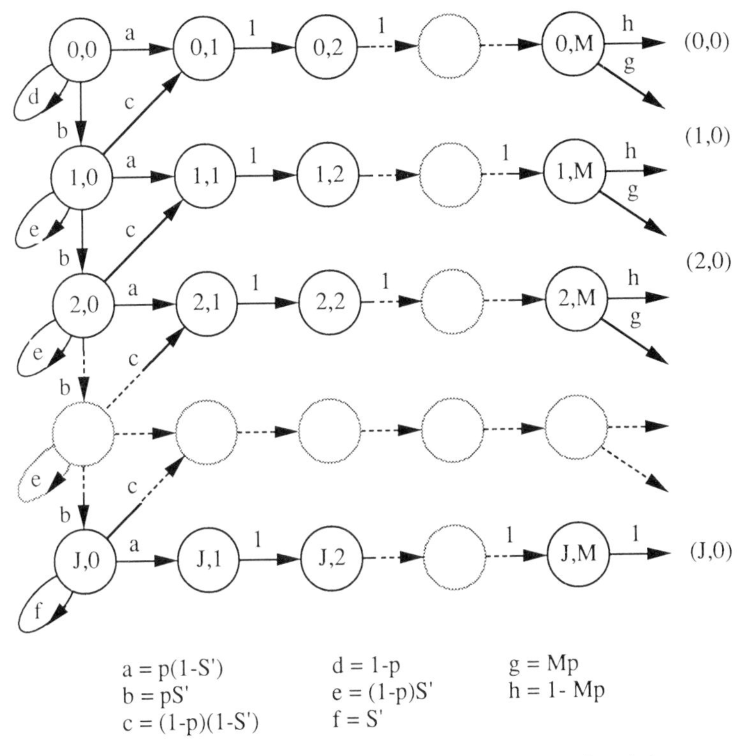


*Figure 7.9 Discrete-time queueing network model for a slotted ring*

A discrete-time queueing network model that can be used to evaluate the performance of the medium access control protocol of the slotted ring just described is given in Fig. 7.9. In formulating this model the following assumptions have been made, in addition to the global assumptions A7.1 to A7.5.

(1) The ring has $N$ users and $M$ slots, where $N > 2$ and $M \leq N$.
(2) Users behave independently and there is a single user class.
(3) Users are spaced at equal intervals around the ring.
(4) Messages consist of single packets ($\gamma = 1$).
(5) Packets are of fixed length and fit exactly into a slot (this implies $H = 1$).
(6) Slots behave independently.
(7) A user generates a packet with probability $p$ per slot, with the packet generation taking place at the start of a slot.
(8) Each user has a finite length buffer of capacity $J$ packets, and FCFS service discipline.
(9) If a packet departs from and a new packet arrives at a users buffer in the same slot, the former is assumed to take place first.

(10) Each user can transmit only one packet at a time.
(11) When a packet gains access to the ring, it is immediately removed from the users buffer.
(12) A user can generate, at most, one new packet during its own transmission period; this packet generation is assumed to occur at the start of the last slot of a transmission period with probability $Mp$.

This last assumption is to simplify the model. The assumption is, in effect, the same as A6.4.5 made in the case of buffered slotted Aloha, and typifies a technique that must often be employed so that models with buffered users are tractable. Its use was justified in Section 6.4. Clearly the assumption restricts the possible values of $p$ to the range $0 \leq p < 1/M$.

Figure 7.9 consists of a two dimensional array of nodes, each labelled $(i, j)$, $0 \leq i \leq J$, $0 \leq j \leq M$, where $i$ denotes that users at node $(i, j)$ have $i$ packets in their buffer, and $j$ denotes the number of slots, including the current one, that a user at node $(i, j)$ has been transmitting the packet at the head of its queue; that is, the elapsed transmission time of users at node $(i, j)$.

In Figure 7.9 $S'$ represents the probability that a user at any of the $(i, 0)$ nodes, $0 \leq i \leq J$ (non-transmitting users), find the current slot full. The transition probabilities between the various nodes are then as shown. For example, users at node $(0, 0)$ are idle and have no waiting packets. They will thus remain at node $(0, 0)$ at the end of the current slot if they don't generate a new packet (probability $1 - p$). They will transmit a packet in the current slot and jump to node $(0, 1)$ at the end of the slot if they generate a new packet and the current slot on the ring is empty (probability $p(1 - S')$). They will jump to node $(1, 0)$ if a new packet is generated and the current slot on the ring is full (probability $pS'$). Once at node $(0, 1)$ a user traverses the sequence of nodes $(0, 1), (0, 2), \ldots, (0, M)$ as its packet circumnavigates around the ring. Each of these nodes acts as a one slot delay. A user at node $(0, M)$ has a packet that has just returned after passing around the ring. This user generates a new packet with probability $Mp$ at the start of the current slot and jumps to node $(1, 0)$ at the end of the slot. If a user at node $(0, M)$ does not generate a new packet (probability $1 - Mp$), it jumps to node $(0, 0)$ at the end of the current slot. The other transition probabilities can be derived in a similar way.

Now let $x^{ij}$ be a random variable representing the number of users at node $(i, j)$, $0 \leq i \leq J$, $0 \leq j \leq M$. The vector

$$x = (x^{ij}: 0 \leq i \leq J, 0 \leq j \leq M) \tag{7.92}$$

is the state vector of the system and can be interpreted as a discrete-time Markov chain in the usual way.

The equilibrium point equations for the system are as follows.

$$x^{00}p = x^{0M}(1 - Mp) \tag{7.93}$$

$$x^{i1}p = x^{i0}p(1 - S') + x^{i+1,0}(1 - p)(1 - S') \qquad i = 0, 1, \ldots, J - 1 \tag{7.94}$$

$$x^{J_1} = x^{J_0} p(1 - S') \tag{7.95}$$

$$x^{i0}[1 - (1 - p)S'] = x^{iM}(1 - Mp) + x^{i-1,M} Mp + x^{i-1,0} pS' \qquad i = 1, 2, \ldots, J - 1 \tag{7.96}$$

$$x^{J0}(1 - S') = x^{JM} + x^{J-1,M} + x^{j-1,0} pS' \tag{7.97}$$

$$x^{ij} = x^{i,j-1} \qquad i = 0, 1, \ldots, J; j = 2, 3, \ldots, M \tag{7.98}$$

In addition to the above, the constraint of a finite number of users on the state space gives the equation

$$N = \sum_{i=0}^{J} \sum_{j=0}^{M} x^{ij} \tag{7.99}$$

Under the above constraint, one of the equilibrium point equations (7.93) to (7.98) will be linearly dependent on the others and will thus not be used in the solution.

Conditional throughput at state $x$ (normalised to packet transmission time) is

$$S(x) = \frac{1}{M} \sum_{i=0}^{J} \sum_{j=1}^{M} x^{ij} \tag{7.100}$$

Using equation (7.98) and (7.99) the above can be expressed in simpler form

$$S(x) = \sum_{i=0}^{J} x^{i1} \tag{7.101}$$

Then the constraint equation (7.99) becomes

$$N = \sum_{i=0}^{J} x^{i0} + S(x)M \tag{7.102}$$

Using equations (7.97) and (7.98) we have

$$x^{J0}(1 - S') = x^{J1} + x^{J-1,1} Mp + x^{J-1,0} pS' \tag{7.103}$$

Now, using equations (7.94), (7.95), (7.96) and (7.103) and writing

$$A = \frac{(1 - p)(1 - Mp)(1 - S')}{p[S'(1 - Mp) + Mp]} \tag{7.104}$$

we find that, in general,

$$x^{J-i,0} = x^{J-i+1,0} A \tag{7.105}$$

and

$$x^{J-i,1} = x^{J-i,0} \frac{p}{1 - Mp} \tag{7.106}$$

both for $i = 1, 2, \ldots, J$.

Using equation (7.105) recursively, then

$$x^{J-i,0} = x^{J0} A^{i} \tag{7.107}$$

for $i = 1, 2, \ldots, J$, which is clearly also valid for $i = 0$. Rearranging this we have

$$x^{i0} = x^{J0} A^{J-i} \tag{7.108}$$

for $i = 0, 1, \ldots, J$.

Next, substituting from equation (7.107) in (7.106) then

$$x^{J-i,1} = x^{J0} \frac{p}{1 - Mp} A^{i} \tag{7.109}$$

for $i = 1, 2, \ldots, J$.

Using equations (7.95) and (7.109) in equation (7.101) and summing the resulting geometric series

$$S(x) = \begin{cases} px^{J0}\left[(1 - S') + \dfrac{A}{1 - Mp}\left(\dfrac{1 - A^{J}}{1 - A}\right)\right] & A \neq 1 \\ px^{J0}\left[(1 - S') + \dfrac{J}{1 - Mp}\right] & A = 1 \end{cases} \tag{7.110}$$

Finally, using equation (7.108) in equation (7.102), and again summing the geometric series

$$N = \begin{cases} S(x)M + \left(\dfrac{1 - A^{J+1}}{1 - A}\right)x^{J0} & A \neq 1 \\ S(x)M + (J + 1)x^{J0} & A = 1 \end{cases} \tag{7.111}$$

By substituting for $x^{J0}$ from equation (7.111) into equation (7.110), the latter becomes a fixed point equation of the form

$$S(x) = f(S(x)) \tag{7.112}$$

This can therefore be solved for $S(x)$ provided we can obtain an expression for $S'$, the probability that a non-transmitting user finds the current slot full. Let a solution be $S(x_e) = S$ which, according to the theory of EPA, is an approximation to the mean throughput $E[S(x)]$. Note that $S'$ is not the same as $E[S(x)]$, the mean (normalised) throughput, in the sense that $S'$ is the mean throughput *as observed by a non-transmitting user*, whereas $E[S(x)]$ is the mean throughput *as observed by a random observer.*

To calculate $S'$ we focus attention on a single user at any of the nodes $(i, 0)$, $0 \leq i \leq J$, (a non-transmitting user), and note that slots passing this user can only be occupied by packets from the other $N - 1$ users. Now let $q$ be the equilibrium probability that a user occupies a slot with one of its packets, and let $X$ be a random variable representing the number of slots occupied (by $N - 1$ users) out of $M$. Then since users are assumed to behave independently, $X$ will have a binomial distribution with parameters $(N-1, q)$, conditioned on the interval $0 \leq X \leq M$. This distribution was discussed in Section 2.4, and has a mean value given by equation (2.162) as

$$E[X] = (N-1)q - \frac{(1-q)(M+1)}{R}\binom{N-1}{M+1}q^{M+1}(1-q)^{N-1-(M+1)} \tag{7.113}$$

where

$$R = \sum_{i=0}^{M}\binom{N-1}{i}q^i(1-q)N^{-1-i}$$

Now $S'$ is given by

$$S' = \frac{E[X]}{M} \tag{7.114}$$

Noting that

$$q = \frac{1}{N}\left[\sum_{i=0}^{J}\sum_{j=0}^{M} x^{ij}\right] \tag{7.115}$$

then from equation (7.100) and the fact that $E[S(x)] \simeq S$, we can write

$$q \simeq \frac{MS}{N} \tag{7.116}$$

An examination of equation (7.113) shows that for practical values of $N$ and $M$ the second term on the right hand side is small, and little accuracy is lost by ignoring it. Thus $S'$ can be approximated as

$$S' = \frac{N-1}{M}q \tag{7.117}$$

which, using equation (7.116) gives

$$S' \simeq \frac{N-1}{N}S \tag{7.118}$$

This is an intuitively satisfying result in that it says that the throughput a user observes due to the other $N-1$ users is simply the total throughput scaled by a factor $(N-1)/N$. In effect, we are approximating a conditional binomial distribution with an unconditioned one on the basis that the right tail of the distribution is small and can be ignored.

Substitution of equation (7.118) into equation (7.110) thus enables the latter to be solved as a fixed point equation for $S(x)$, giving a solution $S(x_e) = S$. Then since throughput is a normalised quantity, the solution we seek is in the interval [0, 1]. It should be noted that expansion of equation (7.110) results in a polynomial of degree $J + 2$ in $S(x)$. That there is a unique root in the interval [0, 1] has not been mathematically proved, although results have always supported this uniqueness. Intuitively, provided the modelled protocol is stable there cannot be more than one value of normalised throughput for a given set of parameters. It follows therefore that there can only be one solution to the fixed point equation over the possible range of throughput, which is the interval [0, 1].

Having found the equilibrium value of normalised throughput, $S$, the mean packet delay, $W$, can be obtained using Little's result by first calculating the mean number of packets stored in the system. Denoting this number by $\bar{n}$, then

$$\bar{n} = \sum_{i=1}^{J} \left( i \sum_{j=0}^{M} x^{ij} \right) + x^{00} p \qquad (7.119)$$

Using equations (7.95), (7.98), (7.105), (7.106), (7.108) and (7.119), after some manipulation we find

$$\bar{n} = \frac{x^{J0}}{1 - Mp} \sum_{i=1}^{J-1} iA^{J-i} + Jx^{J0}[1 + Mp(1 - S')] + x^{J0} A^J p \qquad (7.120)$$

Summing the series, then

$$\bar{n} = \begin{cases} \dfrac{x^{J0}}{1 - Mp} \left[ \dfrac{(A^{J-1} - 1) - (J - 1)(1 - A^{-1})}{(1 - A^{-1})^2} \right] & \\ + Jx^{J0}[1 + Mp(1 - S')] + x^{J0} A^J p & A \neq 1 \\ x^{J0} J \left[ \dfrac{(J - 1)}{2(1 - Mp)} + 1 + \dfrac{p}{J} + Mp(1 - S') \right] & A = 1 \end{cases} \qquad (7.121)$$

Substituting for $S'$ and $x^{J0}$ from equations (7.118) and (7.110), respectively, gives $\bar{n}$ as a function of $S$ (which is now a known quantity). The mean packet delay is then given by

$$W = \frac{\bar{n}}{S} \qquad (7.122)$$

Since this model considers buffered users, the other performance measure of interest is the probability of buffer overflow, $B$. Noting that under the assumptions made a user at node $(J, 0)$ will only lose a packet if the current slot is full, we have

$$B = \frac{1}{N} \left[ x^{J0} S' + \sum_{j=1}^{M} x^{Jj} \right] \qquad (7.123)$$

Using equations (7.98) and (7.118), then

$$B = \frac{x^{J0}}{N}\left[\frac{S(N-1)(1-Mp)}{N} + Mp\right] \tag{7.124}$$

Using equations (7.110) and (7.118) to write $x^{J0}$ as a function of $S$, equation (7.124) gives the equilibrium value of $B$.

Performance results obtained from this model for throughput, mean packet delay, and probability of buffer overflow are given respectively in Figs. 7.10, 7.11 and 7.12, where each quantity is shown plotted against $p$.

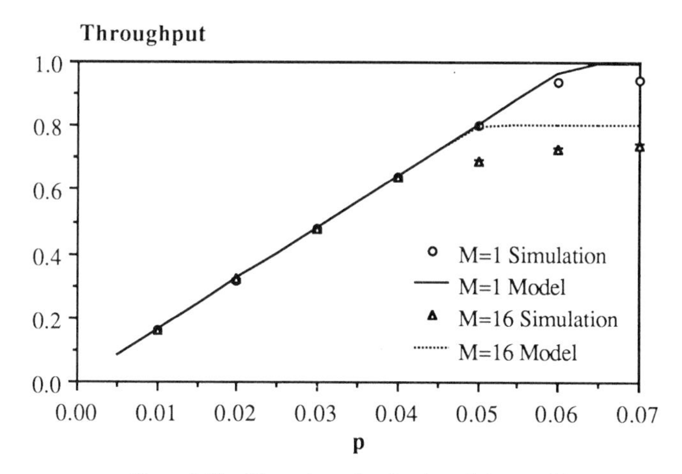


*Figure 7.10 Throughput for the slotted ring model*

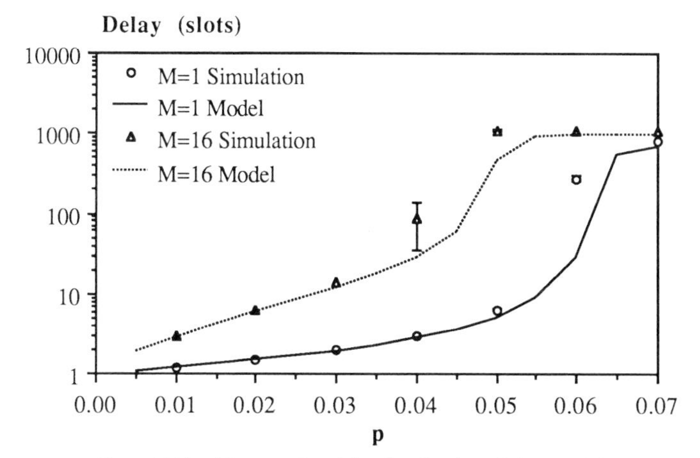


*Figure 7.11 Mean packet delay for the slotted ring model*

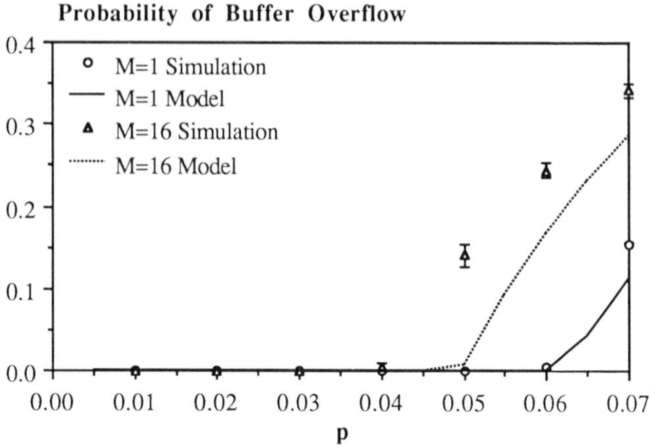

*Figure 7.12 Probability of buffer overflow for the slotted ring model*

The system examined has $N = 16$ users, and results are shown both for $M = 1$ and $M = 16$ slots, these latter two quantities being representative of the two extreme values for the number of slots under the assumption that $M \leq N$. Performance results for intermediate values of $M$ ($1 < M < N$) fall between the two extremes considered. The results are compared with those obtained from a series of $10^5$ slot simulations. The buffer size is $J = 50$ throughout.

In all cases the results obtained from the model with $M = 1$ slot are extremely close to simulation results. With $M = 16$ slots the results are again close at loads up to saturation, after which the model tends to deteriorate and overestimate the ring's performance. This could partly be due to the assumption that was made to simplify the model, that a user cannot generate more than one new packet during its own transmission period. It is likely however that the majority of the error is due to the existence of quasi-stable states, which are known to reduce a slotted ring's transmission capacity. This topic will not be discussed here, but it has been shown in [FALC 85] that such states are prevalent on multiple slot rings at high network loadings, when the probability of the users' buffers being empty is very small.

To take this phenomenon into account it would be necessary to explicitly model the ordering of slots in relation to users on the ring, and it appears very doubtful that any model could take this ordering into account without becoming intractable.

The final performance curve, Fig. 7.13, shows how the model, and indeed any buffered user model, can be used to assess the effects of buffer size on the probability of buffer overflow. Clearly, once the buffer size exceeds a modest value, in this case about $J = 10$ then buffer size has very little effect on the value of $B$ for a given value of $p$. In other words, once the

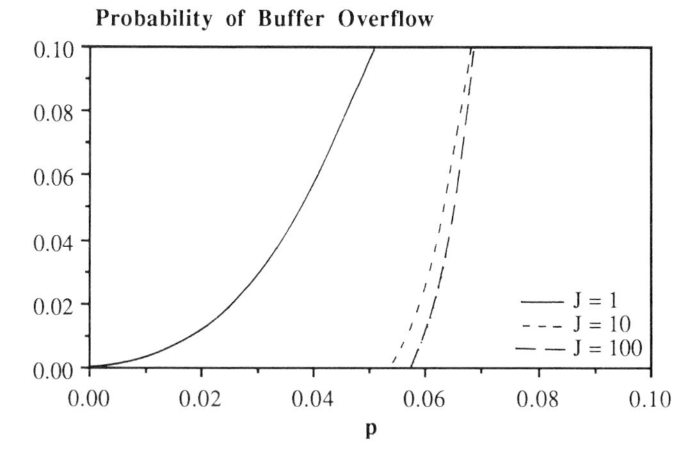


*Figure 7.13 Buffer overflow characteristics for the slotted ring*

rate at which new packets enter the system approaches the maximum rate at which packets can be transmitted, then the queue of packets in a user's buffer will build up very quickly.

Alternative modelling techniques for a slotted ring can be found in [BUX 81], [HARR 85], [KING 87], [KAMA 89] and [AREM 90]. The treatment here follows that of [WOOD 91].

## 7.4 EXERCISES

7.1 Replace the dynamic $p$-persistent backoff algorithm in the CSMA/CD model of Section 7.1 with a 1-persistent binary exponential backoff algorithm. This is commonly known as the Ethernet.
In the Ethernet, a user senses the channel with probability 1 at the $B_i$ nodes, $i = 0, 1, 2, \ldots, m$. If a user is involved in a collision, then after the $K$ slot jamming period a user waits $k$ slots before again sensing the channel, where $k$ is a random integer selected with a uniform distribution from $\{0, 1, \ldots, 2^j - 1\}$, and $j$ is the number of times that a user's packet has collided.
This type of delay can be modelled by the node arrangement shown in Fig. E7.1.

*7.2 Extend the CSMA/CD model of Section 7.1 to allow for buffered users, where each user maintains a buffer that can hold $J$ packets. Single packet messages should be assumed.

*7.3 Extend the token bus model of Section 7.2.1 to allow for buffered users, where each user maintains a buffer that can hold $J$ packets. Single packet messages should be assumed.

7.4 Extend the token bus model of Section 7.2.1 to allow for statistically

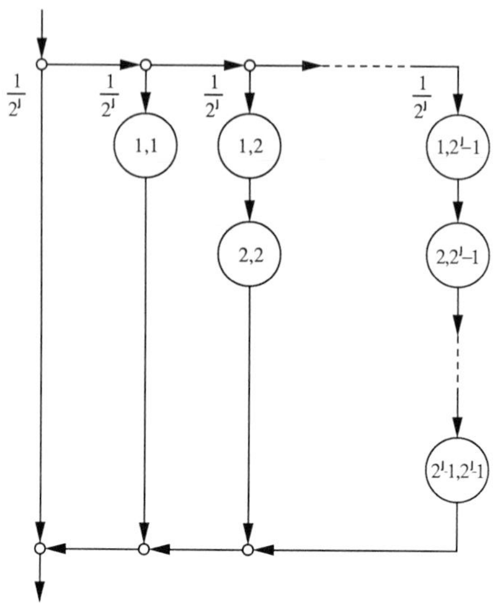


*Figure E7.1 Modelling an Ethernet backoff delay*

different users. This should be done using the REPA algorithm (Section 5.5.1). This model is interesting in that for each increase in the level of the recursion, the number of nodes in the queueing model increases by two! Hint: The key to solving this model is to note that with $K$ users in the network, when a user $i = 1, 2, \ldots, K$ is at either of the nodes $T_j$ or $W_j$, $j = 1, 2, \ldots, K-1$, then user $k$ must be at one of the nodes $T_0$, $W_0$, $S_0$, $S_1, \ldots, S_{H-1}$, $D_0$, $D_1, \ldots, D_{R-1}$, where $k$ is given by

$$k = \begin{cases} K - j + i, & i \leq j \\ (K - j + i) \bmod K, & i > j \end{cases}$$

Then if the state dependent routing probability $T(x)$ associated with node $T_j$ or $W_j$ is denoted by $T_j(x)$, $T_j(x)$ is given by the conditional probability that user $k$ transmits a packet or the token in the current slot, given that user $k$ is at one of the nodes $T_0$, $W_0$, $S_0$, $S_1, \ldots, S_{H-1}$, $D_0$, $D_1, \ldots, D_{R-1}$.

7.5 Modify the timed token protocol model of Section 7.2.2 to allow for pushout priority. This is defined as follows: Pushout priority for class $i = 1, 2$, means that if a class $i$ message arrives while there is a class $j \neq i$ message waiting in the buffer, then the class $j$ message is replaced by the new class $i$ message. Modify the original (non priority) model to give pushout priority to both class 1 and to class 2, and in each case examine how this affects the mean message delay for both classes.

7.6 Modify the slotted ring model of Section 7.3 to allow for statistically different users. This can be done using either REPA or EEPA. The solution to a similar model can be found in [WOOD 93B].

# APPENDIX

Some useful summations of series.

(1) $$\sum_{n=n_1}^{n_2} x^n = \begin{cases} \dfrac{x^{n_1} - x^{n_2+1}}{1-x}, & x \neq 1 \\ n_2 - n_1 + 1, & x = 1 \end{cases}$$

(2) $$\sum_{n=0}^{\infty} x^n = \frac{1}{1-x}, \qquad |x| < 1$$

(3) $$\sum_{n=1}^{\infty} x^n = \frac{x}{1-x}, \qquad |x| < 1$$

(4) $$\sum_{n=n_1}^{\infty} x^n = \frac{x^{n_1}}{1-x}, \qquad |x| < 1$$

(5) $$\sum_{n=1}^{\infty} nx^{n-1} = \frac{1}{(1-x)^2}, \qquad |x| < 1$$

(6) $$\sum_{n=1}^{n_2} nx^n = \frac{x}{(1-x)^2}[1 - x^{n_2}[1 + (1-x)n_2]], \qquad |x| < 1$$

(7) $$\sum_{n=0}^{\infty} \frac{x^n}{x!} = e^x$$

(8) $$\sum_{n=0}^{\infty} \frac{(x \log_e a)^n}{n!} = a^x$$

(9) $$\sum_{n=0}^{n_2} n = \frac{n_2(n_2+1)}{2}$$

(10) $$\sum_{n=0}^{n_2} n^2 = \frac{n_2(n_2+1)(2n_2+1)}{6}$$

(11) $$\sum_{n=0}^{n_2} n^3 = \left[\frac{n_2(n_2+1)}{2}\right]^2$$

# REFERENCES


[AREM 90] van Arem, B., and E.A. van Doorn, 'Analysis of a queueing model for slotted ring networks', van Dijk, N.M., P. Gerrand, W.T. Henderson, R.E. Warfield, and R.G. Addie (Eds.), *Teletraffic Theories for New Telecommunications Services*, Elsevier (North-Holland), 309–314, 1990

[BAKE 81] Baker, C., and C. Phillips (Eds.), *Numerical Solution of Non-linear Problems*, Clarendon Press, Oxford, 1981

[BHAR 80] Bharath-Kumar, K., 'Discrete-time queueing systems and their networks', *IEEE Trans. Commun.*, Com-28, 2, 260–263, 1980

[BHUY 89] Bhuyan, L.N., D. Ghosal, and Q. Yang, 'Approximate analysis of single and multiple ring networks', *IEEE Trans. Comp.*, C-38, 7, 1027–1040, 1989

[BOXM 88] Boxma, O.J., and W.P. Groenendjijk, 'Waiting times in discrete-time cyclic-service systems', *IEEE Trans. Commun.*, COM-36, 2, 164–170, 1988

[BUX 81] Bux, W., 'Local area subnetworks: A performance comparison', *IEEE Trans. Commun.*, COM-29, 10, 1465–1473, 1981

[BUZE 73] Buzen, J.P., 'Computational algorithm for closed queueing networks with exponential servers', *Comm. ACM*, 16, 9, 527–531, 1973

[CHAN 82] Chandy, K.M., and Neuse, D., 'Linearizer: a heuristic algorithm for queueing network models of computing systems', *Comms. ACM*, 25, 2, 126–134, 1982

[CHLA 79] Chlamtac, I., W.R. Franta, and K.D. Levin, 'BRAM: The broadcast recognising access method', *IEEE Trans. Commun.*, COM-27, 1183–1190, 1979

[CONW 89] Conway, A.E., and N.D. Georganas, *Queueing Networks – Exact Computational Algorithms: A Unified Theory Based on Decomposition and Aggregation*, MIT Press, 1989

[COOP 86] Cooper, G.R., and C.D. McGillem, *Modern Communications and Spread Spectrum*, McGraw-Hill, 1986

[CROW 73] Crowther, W., et al., 'A system for broadcast communication: Reservation-Aloha', *Proc. 6th Hawaii Int. Conf. Syst. Sci.*, 371–374, 1973

[DADU 83] Dadunda, H., and R. Schassberger, 'Networks of queues in discrete-time', *Zeitschrift für Operations Research*, 27, 159–175, 1983

[DESM 90] Desmet, E., and G.H. Petit, 'Performance analysis of a discrete-time multiserver queueing system Geo(N)/D/c/K', *Proceedings of RACE Workshop*, Munich, 3–4 July 1990

[DIJK 90] van Dijk, N.M., 'An insensitive product form for discrete-time communication networks', King, P.J.B., I. Mitrani, and R.J. Pooley (Eds.), *Performance '90*, Elsevier (North-Holand), 77–89, 1990

[FALK 85] Falconer, R.M., J.L. Adams, and G.M. Whalley, 'A simulation study of the Cambridge ring with voice traffic', *British Telecom. Tech. J.*, 3, 2, 85–91, 1985

[FARB 72] Farber, D.J., and K.C. Larson, 'The system architecture of the distributed computer system', in Fox, J. (Ed.), *Computer Communications Networks and Teletraffic, Microwave Research Institute Symposia*, 22, 21–27, Polytechnic Press, New York, NY, 1972

[FELL 71] Feller, W., *An Introduction to Probability Theory and its Applications*, Vol. 1 and 2, Wiley, 1971

[FUKU 81] Fukuda, A., and Y. Hasegawa, 'Studies on the URN Scheme for Multiple Access Packet Communication Systems', *Electronics and Commun. in Japan*, 64-B, 10, 56–65, 1981

[GELE 82] Gelenbe, E., and I. Mitrani, 'Control policies in CSMA local area networks: Ethernet controls', *Performance Evaluation Rev.*, 11, 4, 250–258, 1982

[HAMM 86] Hammond, J.L., and P.J.P. O'Reilly, *Performance Analysis of Local Computer Networks*, Addison-Wesley, 1986

[HARR 85] Harrus, G., 'A model for the basic block protocol of the Cambridge ring', *IEEE Trans. Soft. Eng.*, SE-11, 1, 130–136, 1985

[HAYE 74] Hayes, J.F., 'Performance models of an experimental computer communications network', BSTJ, 53, 2, 225–259, 1974

[HAYE 78] Hayes, J.F., 'An adaptive technique for local distribution', *IEEE Trans. Commun.*, COM-26, 1178–1186, 1978

[HAYK 88] Haykin, S., *Digital Communications*, Wiley, 1988

[HEND 90A] Henderson, W., and P.G. Taylor, 'Product form in networks of queues with batch arrivals and batch services', *Queuing Systems: Theory and Applications*, 6, 71–88, 1990

[HEND 90B] Henderson, W., C.E.M. Pearce, P.G. Taylor, and N.M. van Dijk, 'Closed queueing networks with batch services', *Queueing Systems: Theory and Applications*, 6, 59–70, 1990

[HILL 87] Hillier, F.S., and B. Jabbari, 'Analysis of the fixed-assigned TDMA technique with finite buffer capacity', *IEEE Trans. Commun.*, COM-35, 7, 725–729, 1987

[IAZE 84] Iazeolla, G., P.J. Curtois, and A. Hordijk, (Eds.), *Mathematical computer performance and reliability*, 417–428, Elseiver (North-Holland), 1984

[IEEE 85A] *IEEE 802.3 CSMA/CD Access Method and Physical Layer Specifications*, IEEE, 1985

[IEEE 85B] *IEEE 802.4 Token Passing Bus Access Method*, IEEE, 1985

[IEEE 85C] *IEEE 802.3 Token Ring Access Method and Physical Layer Specifications*, IEEE, 1985

[JACK 57] Jackson, J.R., 'Networks of waiting lines', *Oper. Res.*, 5, 518–521, 1957

[JAKO 78] Jacobs, I.M., R. Binder, and E.V. Hoversten, 'General purpose packet satellite networks', *Proc. IEEE*, 66, 1448–1467, 1978

[KAMA 89] Kamal, A.R., and V.C. Hamacher, 'Approximate analysis of non-exhaustive multi-server polling systems with applications to local area networks', *Computer Networks and ISDN Systems*, 17, 1, 15–27, 1989

[KELL 79] Kelly, F.P., *Reversibility and Stochastic Networks*, Wiley, 1979

[KEND 53] Kendall, D.G., 'Stochastic processes occurring in the theory of queues and their analysis by means of the imbedded Markov chain', *Annals of Math. Stat.*, 24, 338–354, 1953

[KING 87] King, P.J.B., and I. Mitrani, 'Modelling a slotted ring local area network', *IEEE Trans. Comp.*, C-36, 5, 554–561, 1987

[KING 90] King, P.J.B., *Computer and Communication Systems Performance Modelling*, Prentice-Hall, 1990

[KLEI 75] Kleinrock, L., *Queueing Systems, Vol. 1: Theory*, Wiley, 1975

[KLEI 78] Kleinrock, L., and Y. Yemini, 'An optimal adaptive scheme for multiple access broadcast communication', *Conf. Rec. IEEE, ICC 78*, 7.2.1–7.2.5, 1978

[KLEI 80] Kleinrock, L., and Y. Yemini, 'Interfering queueing processes in packet-switched broadcast communication', *Proc. IFIP Congress 80*, Tokyo, Japan, and Melbourne, Australia, 557–562, 1980

[KOSO 78] Kosovych, O.A., 'Fixed assignment access technique', *IEEE Trans. Commun.*, COM-26, 9, 1370-1376, 1978

[KUO 81] Kuo, F.F., *Protocols and Techniques for Data Communication Networks*, Prentice-Hall, 1981

[LAM 76] Lam, S.S., 'Delay analysis of a packet-switched TDMA system', *Proc. Nat. Telecommun. Conf.*, 16.3-1–16.3-6, 1976

[LAVE 80] Lavenberg, S.S., and M. Reiser, 'Stationary state probabilities at arrival instants for closed queueing networks with multiple types of customers', *J. Appl. Prob.*, 17, 1048–1061, 1980

[LEE 88] Lee, E.A., and D.G. Messerschmitt, *Digital Communication*, Kluwer Academic Publishers, 1988

[LITT 61] Little, J.D.C., 'A proof for the queueing formula: $L = \lambda W$', *Oper. Res.*, 9, 383–387, 1961

[MART 72] Martin, J., *System Analysis for Data Transmission*, Prentice-Hall, 1972

[MATL 88] Matloff, N.S., *Probability Modelling and Computer Simulation*, PWS-Kent, 1988

[MEIS 58] Meisling, T., 'Discrete-time queueing theory', *Operations Research*, 6, 1, 96–105, 1958

[METC 76] Metcalfe, R.M., and D.R. Boggs, 'Ethernet: Distributed packet switching for local computer networks', *Comms. ACM*, 19, 7, 395–404, 1976

[MITR 87] Mitrani, I., *Modelling of Computer and Communication Systems*, Cambridge, 1987

[NEED 82] Needham, R.M., and A.J. Herbert, *The Cambridge Distributed Computing System*, Addison-Wesley, 1982

[ORTE 70] Ortega, J., and T.W. Rheinboldt, *Iterative Solution of Non-linear Equations in Several Variables*, Academic Press, 1970

[PAPO 84] Papoulis, A., *Probability, Random Variables and Stochastic Processes*, Second Edition, McGraw-Hill, 1984

[PIER 72] Pierce, J.R., 'Network for block switches of data', *Bell Sys. Tech. J.*, 51, 1133–1145, 1972

[PRES 92] Press, W.H., B.P. Flannery, S.A. Teukolsky, and W.T. Vettering, *Numerical Recipes in C*, Second Edition, Cambridge University Press, 1992

[PROA 89] Proakis, J.G., *Digital Communications*, McGraw-Hill, 1989

[PUJO 85] Pujolle, G., J.P. Clande, and D. Seret, 'A discrete tandem queueing system with a product form solution', *Proc. Int. Seminar on Computer Networking and Performance Evaluations*, Elsevier (North-Holland), 139–147, 1985

[PUJO 91] Pujolle, G., 'Discrete-time queueing systems for data networks performance evaluation', in Cohen, J.W., and C.D. Pack (Eds.), *Queueing, Performance and Control in ATM*, 239–244, Elsevier (North-Holland), 1991

[RAYC 81] Raychaudhuri, D., 'Performance analysis of random access packet-switched code division multiple access systems', *IEEE Trans. Commun.*, COM-29, 6, 895–901, 1981

[REIS 75] Reiser, M., and H. Kobayashi, 'Queueing networks with multiple closed chains: Theory and computational algorithms', *IBM J. Res. Dev.*, 19, 3, 282–294, 1975

[REIS 80] Reiser, M., and S.S. Lavenberg, 'Mean value analysis of closed multichain queueing networks', JACM, 27, 313–322, 1980

[REIS 81] Reiser, M., 'Mean value analysis and convolution method for queue dependent servers in closed queueing networks', *Perf. Eval.*, 1, 2, 7–18, 1981

[ROBE 72] Roberts, L.G., 'Aloha packet system, with and without slots and capture', *ARPANET Satellite System Note 8*, (NIC 11290), 1972; reprinted in *Comput. Commun. Rev.*, 5, 28–42, 1975

[ROM 90] Rom, R., and M. Sidi, *Multiple Access Protocols: Performance and Analysis*, Springer-Verlag, 1990

[ROSS 86] Ross, F.E., 'FDDI – a tutorial', *IEEE Communications Magazine*, 24, 5, 10–17, 1986

[ROSS 88] Ross, S., *A First Course in Probability*, Macmillan, 1988

[ROSS 89] Ross, F.E., 'An overview of FDDI: The fibre distributed data interface', *IEEE Journal on Selected Areas in Communications*, 7, 7, 1043–1105, 1989

[SAAD 81] Saadawi, T.N., and A. Ephremides, 'Analysis, stability and optimization of slotted Aloha with a finite number of buffered users', *IEEE Trans. Automatic Control*, AC-26, 680-689, 1981

[SCHW 87] Schwartz, M., 'Modelling and analysis of a token ring', in Cabanel, J.P., G. Pujolle, and A. Danthine (Eds.), *Local Communication Systems: LAN and PBX*, Elsevier (North-Holland), 387–399, 1987

[SEVC 81] Sevcik, K.C., and I. Mitrani, 'The distribution of queueing network states at input and output instants', JACM, 28, 358–371, 1981

[SKLA 88] Sklar, B., *Digital Communications*, Prentice-Hall, 1988

[STAL 85] Stallings, W., *Data and Computer Communications*, Macmillan, 1985

[STAL 87] Stallings, W., *Local Networks: An Introduction*, Second Edition, Macmillan, 1987

[TAKA 86] Takagi, H., *Analysis of Polling Systems*, MIT Press, 1986

[TAKA 90] Takagi, H., 'Effects of target token rotation time on the performance of a timed-token protocol', King, P.J.B., I. Mitrani, and R.J. Pooley (Eds.), *Performance '90*, Elsevier (North-Holland), 363–370, 1990

[TANE 81] Tananbaum, A.S., *Computer Networks*, Prentice-Hall, 1981

[TANE 88] Tananbaum, A.S., *Computer Networks*, Second Edition, Prentice-Hall, 1988

[TASA 86] Tasaka, S., *Performance Analysis of Multiple Access Protocols*, MIT Press, 1986

[TOBA 76A] Tobagi, F.A., and Kleinrock, L., 'On the analysis and simulation of buffered packet radio systems', *Proc. 9th Hawaii Int. Conf. Syst. Sci.*, Univ. of Hawaii, Honolulu, 42–45, 1976

[TOBA 76B] Tobagi, F.A., and L. Kleinrock, 'Packet switching in radio channels: Part III – Polling and (dynamic) split channel reservation multiple access', *IEEE Trans. Commun.*, COM-24, 832–845, 1976

[TOBA 80A] Tobagi, F.A., and V.B. Hunt, 'Performance analysis of carrier sense multiple access with collision detection', *Computer Networks*, 4, 245–259, 1980

[TOBA 80B] Tobagi, F.A., 'Analysis of a two-hop centralized packet radio network: Part I – Slotted Aloha', *IEEE Trans. Commun.*, COM-28, 2, 196–207, 1980

[WALR 83A] Walrand, J., 'A discrete-time queueing network', *J. Appl. Prob.*, 20, 903–909, 1983

[WALR 83B] Walrand, J., 'A probabilistic look at networks of quasi-reversible queues', *IEEE Trans. Inf. Th.*, IT-29, 6, 825–831, 1983

[WALR 88] Walrand, J., *An Introduction to Queueing Networks*, Prentice-Hall, 1988

[WOLF 89] Wolff, R.W., *Stochastic Modelling and the Theory of Queues*, Prentice-Hall, 1989

[WOOD 91] Woodward, M.E., 'Equilibrium point analysis of a slotted ring', Hillston, J., P.J.B. King, and R.J. Pooley (Eds.), *Computer and Telecommunications Performance Engineering*, Springer-Verlag, 150–162, 1991

[WOOD 93A] Woodward, M.E., and K.R.S. Rodrigo, 'Equilibrium point approximations for multiple access protocols with different customer classes', in Schwetman, H., J. Walrand, K. Bagchi, and D. DeGroot (Eds.), *MASCOTS'93*, 299–304, SCS/IEEE/ACM/IFIP, 1993

[WOOD 93B] Woodward, M.E., and K.R.S. Rodrigo, 'Recursive equilibrium point approximations for multiple access protocols', in Bagchi, K., and J. Walrand (Eds.), *MASCOTS'93*, Kluwer Academic Publishers, 1993

# INDEX


Aloha, 134
  reservation, 6
  slotted, 134
    buffered, 154
    delayed first transmission, 69, 115
    different customer classes, 137
    finite channel delay, 73, 117, 141
    unbuffered, 69, 70, 135
    zero channel delay, 134
Aperiodic, 59
  Markov chain, 59
  states, 59
Approximation technique, 113, 114
  EPA, 114
Arrival process, 6, 7
  Bernoulli, 8, 89
  Extended Bernoulli, 98
  Poisson, 49, 89
Arrival theorem, 121
  equilibrium point, 121
Assumptions, 129, 163
  local area networks, 163
  satellite networks, 129
Axioms, 13, 14
  probability theory, 13, 14

Backoff, 163
  algorithm, 163
    dynamic, 164
    binary exponential, 167
Balance equations, 60
  discrete-time $M/M/1$ queue, 80
  discrete-time $M/M/1/J$ queue, 83
  discrete-time $M^{a_n}/M/1$ queue, 86
  Markov chain, 60
  slotted Aloha, 72
    delayed first transmission, 72
Bayes theorem, 17

Bernoulli, 8
  arrival process, 8, 89
  distribution, 30
  trial, 32
  variable, 32
Binary exponential backoff, 167
Binomial distribution, 32, 42
Blocking probability (see probability of blocking)
Bus, 3, 4
  CSMA/CD, 164
  token, 171

Carrier sense multiple access (see CSMA)
Carrier sense multiple access with collision detection (see CSMA/CD)
cdf, 19
CDMA, 149
Central Limit Theorem, 44
Closed queueing network, 104
Code division multiple access (see CDMA)
Complete probability formula, 17
Conditional distribution, 45
  binomial, 47
Conditional probability, 16
Continuous-time, 76
  $M/M/1$ queue, 82
Convolution, 28
  algorithm, 108
Covariance, 27
CSMA, 5, 6
CSMA/CD, 5, 164
Cumulative distribution function (see cdf)
Customer, 7
  class, 120
  population, 7

Delay, 10

Delay *continued*
mean message, 10
mean packet, 193
Discrete-time, 1, 76
Markov chain, 54
queue, 76
$Geo(N)/D/1$, 98
$M/M/1$, 79
$M/M/1/J$, 83
$\boldsymbol{M^{an}/M}/1$, 85
$\boldsymbol{M^{an}/M^{dn}}/\infty$, 90
$\boldsymbol{M/M^{dn}}/c$, 98
$\boldsymbol{M/M^{dn}}/c/J$, 98
$\boldsymbol{M^{an}}/D/1$, 98
S-queue, 93
queueing conventions, 78
queueing networks, 99
Distribution, 33
Bernoulli, 30, 31, 32
binomial, 42
conditional, 45
exponential, 37
extended Bernoulli, 98
geometric, 39
Normal, 44
Poisson, 42
uniform, 33

EEPA, 125
EPA, 114
Equilibrium distribution, 57
Markov chain, 58
Equilibrium point, 114
Equilibrium point analysis (see EPA)
Equilibrium point Arrival Theorem, 121
Equilibrium point equations, 114
Ethernet, 163, 195, 196
Event, 12
Expectation, 24
Exponential distribution, 38
Exponential random variable, 37
Extended Bernoulli distribution, 98
Extended EPA (see EEPA)

Factorial moment, 29
FCFS, 8
FDDI, 163, 178
Feed-forward network, 101
Fibre distributed data interface (see FDDI)
Finite buffer, 154
slotted Aloha, 154
slotted ring, 187
Finite channel delay, 141
slotted Aloha, 117, 141
First-come first-served (see FCFS)
Flow conservation equations, 102

Gaussian (Normal) distribution, 44
Generating function, 27
calculation of expectation, 29
calculation of moments, 29
Geometric distribution, 39

Homogeneous, 54
Markov chain, 54

Independence, 15
Interarrival time, 7
Poisson process, 53
Interfering queues, 116
Invariant distribution, 58
Irreducible Markov chain, 58

Jackson network, 106

Kelly's lemma, 64, 91
Kendall's notation, 7
Kolmogorov's criterion, 68

l'Hospitals rule, 87, 88
Little's result, 11, 77
Local area networks, 2, 3, 162
carrier sensing networks, 164
slotted ring, 186
token passing networks, 171
token bus, 171
token ring, 177
timed token protocol, 178

Markov chain, 53
aperiodic, 59
aperiodic states, 59
homogeneous, 54
irreducible, 58
models, 69
multidimensional, 73
positive recurrent, 60
recurrent non-null, 60
recurrent states, 58
reversed in time, 63
transient states, 58
Markov property, 38, 53
Mean throughput, 10
Mean value analysis (see MVA)
Mean waiting time, 77
Memoryless property, 38, 53
exponential distribution, 38
geometric distribution, 41
Moment generating function, 29
Moments, 24
Multiple-access protocols, 3
MVA, 125, 126

Networks, 1
  communications and computer, 1
    topology, 2
    types, 2
  discrete-time queueing, 99
    closed, 104
    open, 103
Normal distribution, 44
Normalising, 60
  constant, 104
  equation, 60

$O(N)$ notation, 125
One step transition probabilities, 54
  Markov chain, 54
Open discrete-time queueing network, 103

pdf, 20
Performance measures, 10, 77
  discrete-time queue, 77
  multiple-access protocol, 10
Periodic states, 59
  Markov chain, 59
pmf, 22
Poisson process, 49
  decomposition property, 52
  interarrival times, 53
  union property, 51
Power method, 183
Probability density function (see pdf)
Probability distribution, 33
Probability mass function (see pmf)
Probability of blocking, 77
Probability of buffer overflow, 10
Processor sharing (see PS)
Product form theorem, 103, 104
  closed networks of S-queues, 104
  open networks of S-queues, 103
PS, 8, 97

Quasi-reversibility, 95
Queue, 6, 76
  discrete-time, 6, 76
    $Geo(N)/D/1$, 98
    $M/M/1$, 79
    $M/M/1/J$, 83
    $\boldsymbol{M}^{a_n}/\boldsymbol{M}/1$, 85
    $\boldsymbol{M}^{a_n}/\boldsymbol{M}^{d_n}/\infty$, 90
    $\boldsymbol{M}/\boldsymbol{M}^{d_n}/c$, 98
    $\boldsymbol{M}/\boldsymbol{M}^{d_n}/c/\boldsymbol{J}$, 98
    $\boldsymbol{M}^{a_n}/\boldsymbol{D}/1$, 98
    quasi-reversible, 95
    S-queue, 93
Queueing conventions, 78

Random variable, 19

Random variable *continued*
  continuous, 21
  discrete, 22
  mean, 24
  moments, 24
  variance, 25
Recurrent non-null, 60
  Markov chain, 60
  states, 60
Recursive EPA (see REPA)
REPA, 120
Reversibility, 63
  quasi, 95
Ring, 3, 4
  slotted, 186
  token, 177
Routing, 101
  probability, 101

S-queue, 93
  network, 101
    closed, 104
    open, 101
Satellite network, 129
Server, 7
Service discipline, 7
  FCFS, 8
  LCFS, 8
  PS, 8
Service process, 7
Simulation, 35
Slotted Aloha, 134
  buffered, 154
  different customer classes, 137
  finite channel delay, 73, 117, 141
  zero channel delay, 134
State space, 54
State transition diagram, 9
Steady state theorem, 60
  Markov chains, 60
Stochastic matrix, 55
Stochastic process, 49

Tandem S-queues, 100
TDMA, 111, 130
Throughput, 10, 77
Time reversal theorem, 63
  Markov chains, 63
Time reversible Markov chain, 63, 64
Time-division multiple access (see TDMA)
Time-homogeneous Markov chain, 54
Token ring, 177
Traffic equations (see Flow conservation equations)
Transient states, 9
  Markov chain, 58

Transition matrix, 55
  Markov chain, 55
Transition probability, 54
Transition rates, 66
  time reversible Markov chain, 66
Tree, 67

Unbuffered, 69
  S-Aloha, 69, 70, 135
Uniform random variable, 33
  continuous, 33
Uniform random variable *continued*
  discrete, 35
Unit delay, 117

Variance, 25
Virtual token-passing, 6
Visit probability, 58

Waiting room, 7
Waiting time, 77

z-transform, 28